Pascoe Francis Polkinghorne

Zoological Classification

A handy book of reference with tables of the subkingdoms, classes and orders

.

Pascoe Francis Polkinghorne

Zoological Classification
A handy book of reference with tables of the subkingdoms, classes and orders

ISBN/EAN: 9783337274313

Printed in Europe, USA, Canada, Australia, Japan

Cover: Foto ©berggeist007 / pixelio.de

More available books at **www.hansebooks.com**

ZOOLOGICAL CLASSIFICATION:

A HANDY BOOK OF REFERENCE,

WITH

TABLES OF THE

SUBKINGDOMS, CLASSES, ORDERS, &c.

OF THE

ANIMAL KINGDOM,

THEIR CHARACTERS, AND LISTS OF THE

FAMILIES AND PRINCIPAL GENERA.

BY

FRANCIS P. PASCOE, F.L.S. &c.

SECOND EDITION.

WITH ADDITIONS AND A GLOSSARY.

LONDON:

JOHN VAN VOORST, PATERNOSTER ROW.

MDCCCLXXX.

TABLE

OF THE SUBKINGDOMS, CLASSES, AND SUBCLASSES OF THE ANIMAL KINGDOM.

Subkingdoms.	*Classes.*	*Subclasses.*
Protozoa, p. 5	Rhizopoda, 6. Gregarinida, 12. Infusoria, 12.	
Cœlenterata, p. 16 ...	Spongia, 17. Hydrozoa, 20	Hydroida, 20. Discophora, 26. Siphonophora, 28.
	Actinozoa, 30............ Ctenophora, 36.	Zoantharia, 31. Alcyonaria, 34.
	Rhabdophora, 39.	
Echinodermata, p. 40	Crinoidea, 41. Stellerida, 42. Echinoidea, 44. Holothurioidea, 46.	
Vermes, p. 49	Platyelmintha, 50. Nematelmintha, 54. Chætognatha, 57. Gephyrea, 57. Annelida, 58. Rotifera, 63. Polyzoa, 65.	
Arthropoda, p. 70 ...	Crustacea, 70............ Myriopoda, 90. Arachnida, 93. Insecta, 102.	Cirripedia, 71. Epizoa, 74. Entomostraca, 77. Edriophthalma, 82. Podophthalma, 84. Podosomata, 89.
Mollusca, p. 152	Brachiopoda, 154. Lamellibranchiata,155. Pteropoda, 159. Gastropoda, 160. Heteropoda, 167. Cephalopoda, 168. Tunicata, 170.	
Vertebrata, p. 174 ...	Pisces, 175. Amphibia, 192. Reptilia, 197. Aves, 210. Mammalia, 238.	

ADDITIONS AND CORRECTIONS.

Page 38, line 17, *for* Stenosomata *read* Stenostomata.
  „   40, „   22, *for* [ambulacra] *read* [ambulacral feet], and transfer to preceding line after "retractile tube-feet."
  „   49, *after* Nematelmia. *add* Prof. Huxley proposes "to establish a division Trichoscolices," characterized by the presence of cilia, "in order to discriminate the morphological type which they exemplify from those of the Nematoscolices, containing the Nematoidea." With the latter he includes the Nematorhyncha of Bütschli (*Ichthydium* and its allies).
  „   97, line 3, *after* Ixodidæ *add* (Ticks).
  „ 157, „   2 from the bottom, for *Solon* read *Solen.*
  „ 183, „   6 from bottom, *place* Embiotica *before* Ditrema.
  „ 221, „   6 from bottom, *for* Eudynamys *read* Eudynamis.
  „ 246, „   9, *for* HEBEDIDENTATA *read* HEBEDIDENTATI.
  „ 290, „   8 from bottom, *for* an insect *read* a crustacean.

N.B.—The star before the genus denotes that its species are extinct.

PREFACE TO SECOND EDITION.

....

My aim in the first edition of this work was to produce, as its title implies, a handy book of reference to the Classification of the Animal Kingdom, and to bring the contents of the various groups under the eye in the most concise and simple form. Beyond this some general notices were given, and the English names of the species, so far as they had any, and their scientific equivalents. No original remarks were attempted and no opinions expressed, excepting in the synoptical tables, and for them I claimed the indulgence of those who saw their way to a better selection of characters; as to the classification, it is useless not to expect to find differences of opinion. In this edition the latest works have been consulted, especially those of Schmarda and Claus, which, as giving the most recent views of the German naturalists, have been repeatedly referred to. I have not thought it necessary to go into any details respecting divisions, subdivisions, and so on, and their names, which specialists in so many instances delight to produce. Practically they are of little use, and serve chiefly as headings to their author's own pages.

"Nomenclature is so trifling" a subject to the chieftains of science that I hesitate to mention it; but I think it as well to protest here against the barbarous and other objectionable names (sometimes at variance with good taste and even with decency) that have been introduced into science—such, for example, as Battyghur, Butzkopf, Agamachtschich, Know-nothing, Stuff, Jehovah, Cherubim! or such idiotic names, or rather sounds, as Toi-toi, Sing-sing, Giu, Yama-mai, and many others. Indecent names need not be further alluded to. Under the law of priority it is assumed that *any name* must be retained. Surely such a law has its duties as well as its rights. Why should any name be sanctioned that shocks the good taste or feeling of all but the utterly hopeless? This law of priority, too, has turned out to be, as A. Agassiz expresses it, "a mere shuffling of names." It was to have blessed us with a uniform nomenclature; but, under its shelter, names familiar to us for a generation or more are swept away in favour of others published, in some obscure or forgotten work, one or two or twenty years earlier. Were this law to be carried out amongst insects, "a hopeless state of embarrassment" would be the result. In these pages, in the few instances (chiefly amongst birds) where such changes have been made, I have adhered to the familiar name.

I am indebted to my friend Mr. J. W. Dunning, M.A., of Lincoln's Inn, for a long list of errors, in the first edition, in the generic and family names. I have profited by his suggestions in many instances; but, whilst admitting the correct form of the rest, I do not feel quite justified, in a work of this sort, in altering those which have received the sanction, in many cases, of long usage, or have been generally acquiesced in, and which would now be proposed for the first time—such, for example, as *Cheliphorus* for *Chelifer*, *Petromyxon* for *Petromyzon*, *Loliginopsis* for *Loligopsis*. Another class of errors is the non-duplicature of the *r*, as in *Stylorhynchus*, *Biorhiza*, *Ptilorhis*; but this form is all but universal. Again, the practice of making *ma* or *oma* neuter is not so generally adopted that I have thought it necessary to make any change in the termination of the family names when I could find no authority for doing so; *idæ* and *atidæ* are therefore used indifferently. For the omission of that ill-used *h* in such words as *Ryngota*, *Ramphodon*, *Sarcoramphus* I am not responsible, or for its appearance in *Micrhyla*, *Philhydrus*, *Enhydra*, &c. Lastly, objection has been taken to the use as ordinal names of such words as Ecardines, Leptocardii, Polypi, Plectognathi; but here, as well as in others, except where they were intolerable, I have but followed suit. Amongst about 5700 generic names, exclusive of numerous others, mentioned in this little work, there are still, I am afraid, printer's errors and my own to be accounted for.

Out of the 60,000 genera of the animal kingdom I have selected those which are the best known or are the most representative. The genera of the Protozoa, Cœlenterata, and Echinodermata are largely in excess of the other subkingdoms on account of the interest of recent investigations. The class of Insects is least represented in proportion, the Coleoptera especially having less than an eighth of their number.

There are now so many special terms to be met with in Biological science that I have thought a Glossary would be useful.

Some of these terms are only to be met with in what may be called a diffused form, and to give a short and concise definition has been a matter of difficulty. But here, and, as I fear, throughout these pages, I have sacrificed style to brevity. I have inserted many words, such as Evolution, Life, Materialism, &c., and certain anatomical terms which may be an advantage to students.

Since this book has been in the press much has appeared on subjects that I should gladly have availed myself of. Dr. Günther, in that indispensable work the 'Zoological Record' (vol. v.), states that not less than 34,000 pages of "zoological literature" were published within the year 1868. How much has been done within the last six months at home and abroad will only dawn on us by slow degrees; it is now scarcely possible for any one to keep up, except as to the most salient points, with the progress of zoological science. The best will be specialists and a little more. I do not say this to deprecate criticism, which, for the sake of truth, I shall be glad to see freely expressed; it was kindly and, I may say, flatteringly so of the first edition.

F. P. P.

April 27, 1880.

ZOOLOGICAL CLASSIFICATION

&c.

The limits of the Animal Kingdom are still undecided, and it is probable that no absolute division between animals and plants exists. The thin membrane of the animal cell, when present. has been contrasted with the thicker and harder membrane of the vegetable cell, in the former case admitting solid particles of food into the body, as well as enabling it to combine to form the fibrous tissues, which the vegetable cell is unable to do. But it is doubtful if any such hard-and-fast line can be drawn between them.

To evade the difficulty and to exclude all doubtful organisms, Häckel proposes to form an intermediate kingdom, Protista, including Protozoa and the lower plant-forms—Labyrinthuleæ, Diatomaceæ, and Myzomycetes, to which he seems inclined to add the Fungi. But already, in 1859, Owen (Encycl. Brit.) introduced a "kingdom" Protozoa [this name was first proposed by Siebold], including Diatomaceæ &c., which he placed before the "kingdom Animalia;" and John Hogg, the year after, proposed the term "Protoctista" for a "*Primigenal* kingdom," containing both Protophyta and Protozoa. The latter are, however, undoubtedly animal organisms; and in any case the limitation will be an arbitrary one.

It is now considered almost without exception that there are

B

seven fundamental forms or types (subkingdoms or phyla) of Animal life.

They may be tabulated thus:—

No body-cavity .... PROTOZOA.
A body-cavity [*Metazoa*].
  No backbone.
    No intestinal canal............................ CŒLENTERATA.
    An intestinal canal.
      More or less of a radiate structure ...... ECHINODERMATA.
      Structure never radiate.
        With legs, or, if without legs, ver-
          miform.
          Legs never jointed .... VERMES.
          Legs jointed............................. ARTHROPODA.
        Without legs, never vermiform......... MOLLUSCA.
  A backbone .... VERTEBRATA.

Claus adds Tunicata, and it has also been proposed that Spongia should rank as a subkingdom.

The Theory of Descent assumes the common origin of all animals from these seven types; and that these are derived from a single primæval form, this, according to Häckel, " originating by spontaneous generation." The Biogenetic law, that the tribal history of the development of organisms (phylogeny) is represented by their individual development (ontogeny), is assumed to be conclusive of the common origin of all animals.

The primæval form was a Moneron, and its existence is "attested by the fact that the egg-cell of many animals loses its kernel after becoming fructified, and thus relapses to the lower stage of development of a cytod without a kernel, like a Moneron." This Moneron is individually a Monerula, in the second stage becoming an Amœba, individually a nucleated ovulum. Synamœba, a community of Amœbæ, individually a Morula, is the third stage. Planæa, a many-celled primæval animal without a mouth, its ciliated larva a Planula, marks the fourth stage.

Lastly we come to Gastræa, another many-celled primæval animal with intestine and mouth, its larva a Gastrula; and this was the "common primary form of the six higher animal tribes" (Häckel, Hist. Creat., *Eng. transl.*). [The modified diagram below shows the pedigree of the Gastræada, or the animals descended from Gastræa, according to Häckel.] This, the Gastræa-theory as it is called, has been opposed by Salensky, A. Agassiz, Moquin Tandon, and others.

```
                              Vertebrata.
          Arthropoda.              |        Mollusca.
                 |                 |            |
Echinodermata.   |                 |        Polyzoa.
          \      |    Tunicata.    |            |
          Cœlelmintha.    |     Himatega.
                          |
      Spongia.        Scolecida.
          \                   /
          Cœlenterata.   Vermes.
                             |
                        Gastræada.
```

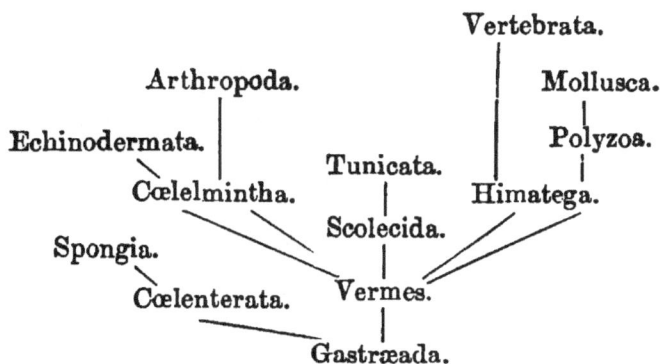

The Himatega or "sac-worms" designate a supposed "stage" of the animal pedigree connecting the Vertebrata with the Invertebrata, whose now nearest relatives are the Ascidians. That the ancestors of man "really existed" in the form of these Himatega "is *distinctly proved*" by the agreement presented by the "ontogeny of *Amphioxus* and Ascidians."

Prof. Huxley, in a paper read before the Linnean Society at the end of 1874, proposed a classification substituting "series for divisions;" for, as he has since remarked, he considers the ordinary mode of arrangement into larger divisions "is a matter of altogether secondary importance."

The following is his tabular arrangement of the animal kingdom :—

"ANIMALIA.

I. *PROTOZOA.*
  i. MONERA.
Protamœbidæ. Protomonadidæ. Myxastridæ. Foraminifera.
  ii. ENDOPLASTICA.
*Amœbidæ. Infusoria flagellata. Gregarinidæ. Acinetidæ.
  ·Infusoria ciliata.   Radiolaria.*

II. *METAZOA.*
  A. GASTREÆ.
    i. POLYSTOMATA.
      Porifera (or *Spongida*).
    ii. MONOSTOMATA.
            1. Archæostomata.
      a. Scolecimorpha.   *b.* Cœlenterata.
Rotifera. Turbellaria.   *Hydrozoa.*
        Trematoda.   *Actinozoa.*
Nematoidea.   Hirudinea.   (*Ctenophora.*)
        Oligochæta.

2. Deuterostomata.

a. Schizocœla.   *b.* Enterocœla.

Annelida *Gephyrea* (?).	*Brachiopoda.*	*Enteropneusta,*
polychæta.	*Polyzoa* (?).	*Chætognatha,*
Arthropoda.	*Mollusca.*	*Echinodermata.*

c. Epicœla.
Tunicata (or *Ascidioida*).
Vertebrata.

B. AGASTREÆ (provisionally).
     Cestoidea.   Acanthocephala."

Journ. Linn. Soc., Zool. xii. p. 226.

Some alterations in this scheme have since been made; Spongida have been added to the type of the Cœlenterata, and the Agastreæ are relegated, the Cestoidea to Trematoda and Acanthocephala to the Nematoidea. The *italicized* groups or series are natural divisions to the extent and limits of which most biologists are agreed.

As the Animal Kingdom can no longer be compared to a "chain extending from the monad up to man," a natural linear arrangement is impossible. From the Protozoa the Cœlenterata branch off on one side and Vermes on the other; from Vermes proceed (1) the Echinodermata, (2) the Mollusca, and (3) the Arthropoda; and from the Mollusca follow, apparently after a long interval, the Vertebrata. The annexed scheme will show at once their relative position:—

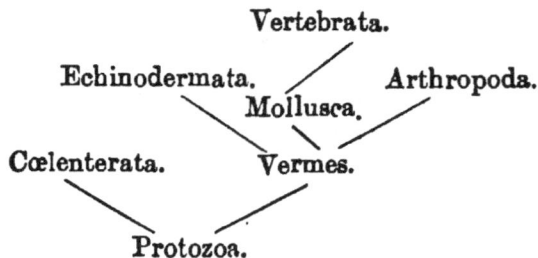

Subkingdom I. **PROTOZOA.**

ACRITA. AMORPHOZOA. SARCODEA. PLASTIDOZOA. HYPOZOA.

Minute, nearly structureless animals, composed of a gelatinous matter [protoplasm or sarcode], and not developing true layers. Reproduction principally by fission, or by the breaking up of the nucleus.

The Protozoa differ from the rest of the animal kingdom in that they present no structural elements, or, even if they possess distinct cells, these cells do not develop into tissues. Such tissues as may be found in the higher Infusoria originate, not from the cells, but by changes in the physical and chemical characters. It is, however, undecided whether all Protozoa are unicellular.

In Monera there is no nucleus; and it has only been recently recognized in some of the Foraminifera. In all other Protozoa there is a nucleus [=endoplast, Huxley].

In the absence of a nervous system, and in its inherent self-acting power, Bowerbank suggests the hypothesis that the sarcode may be a diffused form of nervous matter.

One mode of reproduction is by conjugation (zygosis). Two bodies come together, and a fusion more or less complete takes place. After a time the nucleus breaks up into a number of spores, or the spores are emitted in clouds without any apparent rupture of the surface.

Bathybius, supposed to have been a living protoplasmic substance, is now known to be "little more than sulphate of lime precipitated in a flocculent state by strong alcohol."

The classification and even the limits of the Protozoa are still contested; and the descriptions of these organisms are "in many instances very contradictory." It is doubtful whether many of them can be regarded as any thing more than stages in the development of other animals or of plants. Claus (1876) has two classes —Rhizopoda and Infusoria—treating the remainder as outside the animal kingdom, and more related to Algæ and Fungi. These are Schizomycetes (*Bacteria*), Myxomycetes (*Trichia, Æthalium*); Flagellata (*Monas, Volvox, Euglena, Peridium, Noctiluca*); Catallacta (*Magosphæra*); Labyrinthuleæ, apparently related to the Diatomaceæ; and Gregarinida. Schmarda (1877) has five classes—Rhizopoda, restricted to Amœboidea, Acinetidæ, and Foraminifera ; Polycystina (=Radiolaria); Gregarinoidea ; Infusoria; and Spongia. In Infusoria he places *Vibrio, Bacterium*, and *Volvox*; his order Cymozoida includes the first two, while *Volvox*, together with *Monas, Astasia, Noctiluca*, &c., are comprised under Mastigophora. Huxley divides the Protozoa into Monera (including Foraminifera) and Endoplastica (including Protoplasta (Amœba, &c.), Gregarinida, Infusoria, &c.).

For the present three classes may be recognized :—

Without a mouth.
    Without pseudopodia .... GREGARINIDA.
    With pseudopodia .... RHIZOPODA.
With a mouth .... INFUSORIA.

Class I. RHIZOPODA.

SARCODINA.

Minute aquatic animals, moving by an extension of their substance (pseudopodia); with or without a shell, and without a mouth.

The pseudopodia or extending portions of the substance are sometimes confined to one side of the body; they may be merely short lobular dilations, or, as is generally the case, long filamentous processes capable of multiplying themselves and coalescing whenever they come in contact.

The *nucleus* is a hyaline vesicle generally containing a bluish nucleolus ; the *central capsule* is a membranous sac separating the nucleus from the outer protoplasm [sarcode]. The latter contains granules and certain yellow cells [sarcoblasts] supposed to be parasitic organisms (Cienkowski), an incipient form of liver (Häckel), and reproductive organs (Wallich). Recent observations seem to show that the "reproductive cells" are contained in the central capsule ; and the yellow cells are found by Häckel to contain a substance which cannot well be distinguished from the starch of plants. According to Sir W. Thomson, the capsule is "absent, or, at all events, exists in a very modified form in the more typical groups."

Contractile vacuoles are certain cavities having a rhythmical movement: their function is possibly respiration.

Four orders are indicated, but exceptions occur to most of their characters :—

Without a shell.
    Without a nucleus .... MONERA.
    With a nucleus .... AMŒBOIDEA.
With a shell.
    · Shell calcareous .... FORAMINIFERA.
    Shell siliceous .... RADIOLARIA.

Order I. MONERA.

Homogeneous, structureless, naked particles of albumen capable of nourishment and reproduction. No nucleus nor contractile vesicle.

In their mode of feeding and development, the Monera agree with the Foraminifera. Several of the forms are subject to becoming encysted, and then to breaking up into spores, which may or may not at first resemble the parent. Others, assuming an amœboid condition, may, when two come in contact, unite their pseudopodia and form a contractile network [plasmodium]. In *Protomyxa* the spores unite to form a body like the parent.

There are about fifteen species known, which are found in fresh water as well as in the sea.

There are two divisions :—

Naked (never encysted), reproduced by division... GYMNOMONERA.
Encysted in a structureless membrane during a
    quiescent stage, then breaking up into spores ... LEPOMONERA.

GYMNOMONERA.	LEPOMONERA.
Protamœba.	Protomonas = Monas.
Protogenes.	Protomyxa.
Myxodictyum.	Vampyrella.
	Myxastrum.

Order II. AMŒBOIDEA.

AMŒBINA. LOBOSA. ATRICHA. PROTOPLASTA. SPHYGMICA.

Homogeneous, nearly structureless animals, with nucleus and contractile vesicles. Pseudopodia mostly short and broad, neither ramifying nor coalescing.

These are mostly freshwater organisms, although a few are found in moss or in the earth; they resemble the colourless blood-cells of man, from which they cannot be distinguished; but it is doubtful whether many of them can be any thing more than the earlier stages of other organisms. A sort of carapace or discoid shell is common to *Arcella*; in *Difflugia* it is replaced by a shell formed by foreign particles agglutinated together.

The limits of this order are undecided. Schmarda unites Amœbidæ with Acinetidæ to form his order " Thalamia, or Rhizopoda nuda." Von Hayek includes in it the three families of Amœbidæ, Arcellidæ, and Actinophryidæ; Claus places the latter in the Radiolaria, but Häckel, followed by Mivart, excludes it.

The Heliozoa are mostly freshwater organisms, sometimes provided with a siliceous skeleton of detached radiating spines, or with a hollow perforated globular shell (*Clathrula*). Except Actinophryidæ, they appear more nearly related to Radiolaria than to Amœboidea.

Pelomyxa is an "amœboid organism" "spreading over the bottom of stagnant pools."

The normal Amœboidea include two families :—

Amœbidæ.	*Arcellidæ.*
Amœba.	Arcella.
Corycia.	Euglypha.
Podostoma.	Difflugia.

The Heliozoa include three families. *Actinophrys sol* is the well-known "sun-animalcule" of microscopists. Actinophryidæ constitute the order Phlœophora of Carus. *Magosphæra* is remarkable in that in its life-history it presents four independent forms; the first morphologically represents an egg, the second is a *Volvox*-form, the third is a ciliated Infusorian, and the last apparently an *Amœba*. It was discovered on the Norwegian coast by Häckel, who considers it the representative of a distinct group (Catallacta) linking together several of his Protista; but Huxley thinks it should form a "subdivision" of Infusoria ciliata.

Actinophryidæ.	*Acanthocystidæ.*	*Clathrulinidæ.*
Actinophrys.	Acanthocystis.	Clathrulina.
Actinospbærium.	Rhaphidophrys.	Hyalolampe.
Ciliophrys.	Heteroplrys.	Hedriocystis.
	Cystophrys.	

Magosphæra.

Order III. FORAMINIFERA.

POLYTHALAMIA. RETICULARIA. ACYTTARIA. HOMOGENEA.
THALAMOPHORA.

Homogeneous, nearly structureless animals, without a central capsule or a contractile vacuole, generally provided with a partially enclosed calcareous shell, with or without a nucleus (or nuclei).

Locomotion is performed by extending portions of the surface; these are capable of multiplying themselves and coalescing whenever they come in contact. These pseudopodia also act as prehensile organs. The shell may be either simple or divided by septa into chambers, and is often of extraordinary complexity; occasionally, as in *Lituolidæ*, it is partially composed of fine sand agglutinated together. In *Hyperammina* it is almost entirely composed of silica. The shells are extremely liable to vary, according to age and locality: the more simple have only the terminal aperture [Imperforata]; in the Foraminifera proper [Perforata] the shell is pierced by numerous pores, through which the pseudopodia are emitted.

The complex forms of their shells are said by Häckel to be "traceable to the active agency of the formless albuminous combinations of protoplasm," the "results of *inherited adaptation*."

The Foraminifera are marine, and are amongst the earliest forms of life; the chalk formations and nummulitic limestones are almost entirely composed of their shells, and much of the Atlantic bed is covered with an ooze largely composed of *Globigerina*-shells, which, beyond a depth of 2400 fathoms, are supposed to pass gradually by their decomposition into the red clay which covers many thousand square miles of its bottom. In a living state it is believed they may be found at all depths. They are mostly very minute.

Eozoon is the oldest known fossil. *Parkeria*, another fossil, is comparatively of large size (three inches in circumference). *Rupertia* is a sessile form from the Greenland seas. Many forms, however, are now said to be Algæ (e. g. *Ovulites, Dactylopora*, &c.).

IMPERFORATA.	*Miliolidæ.*	*Lituolidæ.*
Gromiidæ.	Miliola.	Hyperammina.
	Orbitulites.	Spirillina.
Gromia.	Cornuspira.	Lituola.
Lieberkuhnia.		*Parkeria.

PERFORATA.

Lagenidæ.
Orbulina.
*Lagena.

Nodosariidæ.
Nodosaria.
Dentalina.

Soritidæ.
Sorites.

Globigerinidæ.
Rotalia.
Uvellina.
Textularia.
Globigerina.
Polytrema.
Carpenteria.

Rupertia.

Nummulinidæ.
Cristellaria.
Nonionina.
Nummulina.
Polystomella.
Acervulina.
*Nummulites.
*Eozoon.

Order IV. RADIOLARIA.

POLYCYSTINA.   CYTOPHORA.

Body differentiated into ectosarc and endosarc, and provided with one central capsule or with many [nuclei]. Pseudopodia rod-like, radiating, little disposed to ramify or coalesce. Shells siliceous, external.

The shells are more or less perforated for the emission of the pseudopodia, and often furnished with radiating spicules; they are of great variety of form and beauty, and of remarkable complexity. They are all oceanic, and, when alive, are of the most brilliant colours; they appear only to come to the surface after sunset; but they are believed to exist at all depths in the sea. In the fossil state they largely contribute to form the Tertiary rocks.

A group of extremely minute forms, "approaching, but in many important points differing from, the Radiolarians," has been brought to light by the 'Challenger' expedition. They have received the ordinal name of "Challengerida." They have monothalamous siliceous shells, richly sculptured and filled with a nucleated sarcode. Sir Wyville Thomson considers their position zoologically "not very far from such forms as *Gromia.*"

Claus divides the Radiolaria into five suborders:—Heliozoa [here referred to Amœboidea], Thalassicollea, Polycystinea, Acanthometræ, and Polycyttaria. They have also been divided into Collozoa (numerous nuclei) and Collida (a single nucleus). Häckel has two sections—Monozoa and Polyzoa; the former he divides into Ectolithia (skeleton external to the cap-

sule) and Entolithia (skeleton more or less within the capsule).
Mivart proposes to divide these into seven " primary groups "—
Discida, Flagellifera, Entosphærida, Acanthometrida, Polycys-
tina, Collozoa, and Vesiculata.

"These beautiful symmetrical and complex forms cannot be
due to the action of *natural selection* ; and *sexual selection* can of
course take no part in forming such organisms as these. We
seem here to have forced upon our notice the action of a kind of
organic crystallization,—the expression of some as yet unknown
law of animal organization, here acting untrammelled by adaptive modification, or by those needs which seem to be so readily
responded to by the wonderful plasticity of the animal world."—
Mivart.

Having a single central capsule ... MONOCYTTARIA (or MONOZOA).
Having several central capsules ... POLYCYTTARIA (or POLYZOA).

MONOCYTTARIA.

Thalassicollea.

Thalassicollidæ.
Thalassicolla.
Myxobrachia.

Thalassosphæridæ.
Thalassosphæra.
Physematium.

Aulacanthidæ.
Aulacantha.

Acanthodesmiidæ.
Acanthodesmia.
Dictyocha.

Polycystina.

Eucyrtidiidæ.
Eucyrtidium.
Litharachnium.
Dictyopodium.
Dictyoceras.

Ethmosphæridæ.
Ethmosphæra.

Heliosphæridæ.
Heliosphæra.

Aulosphæridæ.
Aulosphæra.

Acanthometræ.

Acanthometridæ.
Acanthometra.
Amphilonche.
Lithoptera.

Cœlodendridæ.
Cœlodendron.

Cladococcidæ.
Cladococcus.

Diploconidæ.
Diploconus.

Haliommatidæ.
Haliomma.
Actinomma.
Dorataspis.

Sponguridæ.
Spongurus.
Spongosphæra.

Coccodiscidæ.
Coccodiscus.
Euchitonia.

Litheliidæ.
Lithelius.

POLYCYTTARIA.
Sphærozoidæ.
Sphærozoum.

Collosphæridæ.
Collosphæra.

----

* Traquairia.

Class II. GREGARINIDA.

One- or two-celled organisms, with a nucleus, and sometimes a nucleolus, but no contractile vesicle. No pseudopodia.

Mostly small organisms which resemble, and have been mistaken for, ova. They differ from single-celled plants in that their tissues are contractile and soluble in acetic acid. Sometimes one extremity is furnished with hooks, serving as organs of prehension. They are all internal parasites; they differ from Amœbæ by having an external cell-membrane.

A conjugative process, so-called, sometimes occurs : two Gregarinida come together, a cyst forms around them, and certain globular vessels are formed in it, which ultimately become peculiar bodies known as "pseudo-navicellæ," or "pseudo-naviculæ." After a time these escape, and, bursting, give rise to amœbiform bodies which develop into Gregarinida.

In the perfect form these parasites occur principally in insects, crabs, and worms. They have been also found in the human heart, kidney, &c. They vary in size ; some attain the length of half an inch.

Monocystidæ.
Monocystis.

Dicystidæ.
Gregarina.
Pixinia.

Stylorhynchus.
Actinocephalus.
Hoplorhynchus.

Didymophyidæ.
Didymophyes.

Class III. INFUSORIA.

POLYGASTRICA. MICROZOA.

Minute, aquatic animals, or occasionally internal parasites of definite form, swimming by the vibration of cilia. A mouth and rudimentary stomach. No pseudopodia. A contractile vesicle and nucleus.

An oral aperture or mouth is mostly confined to the ciliated Infusoria ; and an aboral aperture is sometimes present. The other Infusoria are somewhat doubtful organisms. The Acinetidæ have been regarded as the ancestral forms of the true Infusoria.

The ciliated Infusoria are composed of a cuticle lined with a layer of gelatinous matter [endosarc] filled with chyme or a semi-fluid substance, within which the particles of food rotate. Besides the cilia, many Infusoria are furnished with bristles, hooks, &c., and some with thread-cells [trichocysts].

In most Infusoria, as well as in *Amœba* and others, certain clear spaces or contractile vesicles exist. Other clear spaces [vacuoles] sometimes make their appearance in every part of the body. They vary in size, number, and position.

A few species are found in the sea; others are intestinal (Opalinidæ, *Belantidium, Plagiostoma,* &c.). Opalinidæ were supposed to be the earlier stages of the Trematode worms.

Reproduction is by fusion or by gemmation, as well as by the nucleus at certain periods breaking up into fragments, each developing into the parental form.

Without retractile tubes.
Moving by cilia .... CILIATA.

Moving by flexible filaments .... FLAGELLATA.
With retractile tubes .... SUCTORIA.

Order I. CILIATA.

STOMATODA.

Body more or less provided with vibratile cilia. No filaments [flagella].

The cilia, which are organs of prehension as well as of motion, are expansions of the cuticle [exoplasm]. In the majority of the Infusoria, however, the cuticle is "nothing but a lifeless exudation of the surface;" where it exists it is seen under various forms: in *Dictyocysta* it assumes the form of "lattice-like" shells; in *Vaginicola* it is a protective sheath.

Some species, as in *Vorticella,* are attached to foreign bodies by a long and contractile pedicle. *Torquatella* is a marine Infusorium without cilia; Ray Lankester regards it as a section of Ciliata which he calls Calycata.

The great majority of the Infusoria belong to this order, which has been divided into four suborders or sections dependent on the position of the cilia: in Holotricha they are dispersed, in Heterotricha a longer series is found near the mouth, in Hypotricha they are confined to the under surface of the body, and in Peritricha they are arranged round the mouth.

Omitting Acinetidæ, Claus's arrangement of the families is here followed.

HOLOTRICHA.

Opalinidæ.
Opalina.

Tracheliidæ.
Trachelius.
Loxodes.

Enchelyidæ.
Enchelys.
Leucophrys.
Coleps.
Lacrymaria.

Paramœciidæ.
Paramæcium.
Colpoda.
Nassula.

Cinetochilidæ.
Leucophrys.
Ophryoglena.
Cinetochilum.
Trichoda.
Cyclidium.

HETEROTRICHA.
Bursariidæ.
Plagiotoma.

Balantidium.
Bursaria.

Stentoridæ.
Stentor.
Freia.

Spirostomidæ.
Spirostomum.
Blepharisma.

HYPOTRICHA.
Chlamydodontidæ.
Chlamydodon.
Chilodon.
Huxleya.

Aspidiscidæ.
Aspidisca.

Euplotidæ.
Euplotes.
Uronychia.

Oxytrichidæ.
Oxytricha.
Onychodromus.
Cerona.

———
Torquatella.

PERITRICHA.

Halteriidæ.
Halteria.

Tintinnidæ.
Tintinnus.
Codonella.
Dictyocysta.

Trichodinidæ.
Trichodina.
Urceolaria.
Trichodinopsis.
Didinium.

Vorticellidæ.
Vorticella.
Epistylis.
Gerda.
Carchesium.
Vaginicola.
Ophrydium.
Lageuophrys.
Spirachona.

Ophryoscolecidæ.
Ophryoscolex.
Entodinium.

Order II. FLAGELLATA.

MASTIGOPHORA.

Body provided at its anterior portion with one or more long flexible filaments [flagella], the ends only vibratile. Cilia occasionally present.

The flagella are never more than ten; sometimes there is only one, more frequently two. They serve for locomotion. *Astasia* has a terminal mouth. Fission seems to be the only mode of ⸢ reproduction. These organisms sometimes occur in colonies. ⸥ *Lophomonas* is found in the intestines of *Periplaneta orientalis*.

Euglenæ occur in green masses in our ponds in the spring. Flagellated organisms are sometimes found in the blood. Some are scarcely distinguishable from the spores of certain Algæ. *Peridinium* is phosphorescent.

Noctiluca, forming the class Myxocystodea of Carus, is a globular organism furnished with a short stalk, a mouth, and a digestive cavity; in temperate climates it is the principal cause of the luminosity of the sea. According to Allman, the special seat of the phosphorescence is the peripheral layer of protoplasm which lines the exterior structureless membrane.

Peridiniidæ.	*Astasiidæ.*	*Monadidæ.*
Peridinium.	Euglena.	Monas.
Ceratium.	Astasia.	Cercomonas.
	———	Lophomonas.
	Noctiluca.	

Order III. SUCTORIA.

ACINETA-forms. POLYSTOMA. TENTACULIFERA.

Body stalked, provided with radiating retractile tubes, having at the extremity a disk-shaped mouth, acting as a sucker. No cilia in the adult.

The prey is seized by these tubes, and the nutrient matter is then imbibed by the suckers. They have no definite mouth. There is only one family:—

Acinetidæ.

Acineta.
Podophrya.
Ophryodendron.

A remarkable group of Infusoria of very minute size, distinguished by the presence of a hyaline "wineglass-shaped structure" at the anterior extremity, from which the flagellum takes its origin, has been investigated by J. Clark, and more recently by Saville Kent and Bütschli. They have been called "Flagellate" or "Collar-bearing Monads." The principal genera are *Codosiga, Salpingœca, Codomœca, Dinobryon,* and *Antophysa.*

Subkingdom II. CŒLENTERATA.

RADIATA. ZOOPHYTA. CENTRONIÆ.

Aquatic animals with a distinct body-cavity, and a mouth opening into it; no intestinal canal. Reproduction normally by spermatozoa and ova, which are discharged through the mouth.

The majority of the Cœlenterata are composite animals—that is, organisms made up of colonies or communities of individuals organically united. The substance of the body, which is mostly of a radiate character, is composed of two membranes—ectoderm and entoderm. There are no traces of a nervous system, except in certain Medusæ, and there is no proper blood-vascular apparatus, although there is a fluid in the body-cavity which " represents " the blood.

Peculiar stinging-organs, supposed to be poisonous, are invariably present, except in Sponges; they are called "cnidæ," "nematocysts," or "thread-cells;" they are usually colourless, elastic, double-walled sacs, round or oval, with a fluid in their interior. The inner wall of the sac is produced into a sheath terminating in a long thread [ecthoreum]; this is usually twisted in many irregular coils round its sheath, and fills up the open end of the sac. "Under pressure or irritation, the cnida suddenly breaks, its fluid escapes, and the delicate thread [cnidocil] is projected, still remaining attached to its sheath." The cnidæ are said to be analogous to the tactile organs of the Arthropoda.

The body-cavity of the Cœlenterata does not, according to Häckel, represent the true intestinal cavity [cœloma] of the higher animals, but, in many, a system of cavities [enterocœle] takes its place.

There are four classes :—

Fixed, or, if free, not moving by means of cilia.
   No urticating organs.. SPONGIA.
   With urticating organs.
      Digestive cavity continuous with the body-
         cavity .... HYDROZOA.
      Digestive cavity separated from the body-cavity ACTINOZOA.
Free, swimming by means of cilia .... CTENOPHORA.

Class I. SPONGIA. (Sponges.)

AMORPHOZOA. PORIFERA. POLYSTOMATA. SPONGOZOA.

Fixed aquatic organisms, composed of an aggregate of amoebiform bodies, each provided with a mouth and numerous pores, and including a fibrous framework, strengthened by horny or calcareous spicules. Larvæ free-swimming.

The gelatinous sarcode forming the animal mass of the sponge is made up of a number of individual masses of protoplasm [plastides], forming a thin outer layer and entering deeply into the organism, coating every cavity in the interior. The cavities are connected by canals, which are continuous with ciliated chambers. Minute chambers [pores or ostioles], through which the water passes into the cavities, exist on the outer layer, and the water so admitted is discharged by larger orifices [oscula]. It is to the vibratile action of the cilia that the circulation of the water is due. According to Huxley, the "sponge represents a kind of subaqueous city, where the people are arranged about the streets and roads in such a manner that each can easily appropriate his food from the water as it passes along."

It is doubtful, however, whether the digestion is carried on by the general cells lining the passages or by the ciliated cells.

By Leuckart and Häckel the canal-system of Sponges is regarded as homologous to the gastrovascular system of the Hydrozoa and the Actinozoa. Furthermore, the Sponges and Corals are regarded as blood-relations, both originating [hypothetically] from a primitive sac [protosaccus], the only morphological character separating them being the absence of urticating organs in the former.

Thread-cells are, however, said to be found by Eimer in *Reniera* [these are by Carter declared to be parasitical polyps]. Eimer also considers that he has established a passage between Sponges and Hydroids. By Carter the relation of Sponges to Ascidians is regarded as greater than to Corals, the latter having only one aperture.

Reproduction is either asexual by budding &c., or, in the Calcispongiæ principally, by ova. According to Huxley, the embryo is "similar to the corresponding stage of a hydrozoon, and is totally unlike any known condition of a protozoon." The ova, so called by some observers, are supposed by Häckel to be spermatozoa, or perhaps vibratile cells. Sponges are also reproduced by gemmation. The Gastrula stage is disputed by Barrois and Hyatt.

According to Saville Kent, "Sponges are compound, colony-building, collar-bearing, flagellate monads, exhibiting neither in

C

their embryological nor in their adult condition phenomena that do not find their parallel among the simple unicellular Protozoa."

The skeleton or internal framework of sponges is strengthened either by calcareous, siliceous, or ceratose spicules of various forms; and a simple classification into three orders has been founded on this character. It sometimes happens that a sponge, as *Dysidea*, forms for itself a skeleton of spicules of other sponges or other foreign substances.

The Physemaria of Häckel are supposed to be sponges which do not go beyond the Gastrula stage; they have no pores and are fixed. Carter and Saville Kent assert their foraminiferal nature. Recently Ray Lankester decides *Haliphysema* to be an *Amœba* enclosed in a test of sponge-spicules. Norman forms of it an order which he calls Psammoteichina.

The common sponge of commerce is *Spongia officinalis*; the freshwater sponge, *Spongilla fluviatilis*; Neptune's cup, *Raphiophora patera*.

Almost every one who writes on Sponges has a classification of his own. Gray had at least three, the most elaborate (though not the latest) being marked by the excessive multiplication of genera, which, as he himself observes, are "founded on very different principles and characters" by different authors. Carter has divided the Sponges into eight orders, including numerous families and groups, the latter with hybrid and bizarre names, Bowerbank into three, Schmarda into nine, and Claus into two. Häckel has three legions—one of these, Calcispongiæ, is the subject of a remarkable work. His method obliges him to sweep away the old genera and to create new ones, whose names are drawn successively by affixes from the representatives of his three orders—Ascones, Leucones, and Sycones. These are characterized as "spurious genera" by Norman, who observes that our common *Grantia compressa*, with its varieties and "possible modifications," has 28 generic, subgeneric, and subspecific names, which might be further extended to 54. But all sponges are from their "unlimited pliability" subject to perpetual variation, and sometimes different form-species arise out of "one and the same stock," "which, according to the usual system, would belong to several quite distinct genera."

The Calcispongiæ Häckel affirms, "with the greatest certainty," were developed from *Olynthus*. The "order Ascones" was the first to develop, from which the Leucones and Sycones arose as "diverging branches."

The British species of calcareous sponges belong, according to Häckel's nomenclature, to the following genera:—*Grantia com-*

pressa and *G. ciliata* to *Sycandra*, also *G. ensata* (*S. glabra*) and *G. tessellata* (*S. elegans* = *Dunstervilla*); *Leucosolenia botryoides* to *Ascaltis*; *L. contorta* to *Ascandra*; *L. lacunosa* to *Ascortis*; *L. coriacea* to *Ascetta*; *Leuconia nivea*, *L. fistulosa*, and *Leucogypsia Gossei* to *Leucandra*; and *Leuconia pumila* to *Leucaltis*.

Following Häckel, and referring the Silicispongiæ, Ceratospongiæ, and Hyalospongiæ to the Fibrospongiæ, we have three orders :—

Gelatinous, no skeleton.............................. MYXOSPONGIÆ.
With a skeleton.
    Skeleton siliceous or horny .... FIBROSPONGIÆ.
    Skeleton calcareous .... CALCISPONGIÆ.

MYXOSPONGIÆ.
Halisarcidæ.
Halisarca.

Chondrillidæ.
Gummina.
Chondrilla.

FIBROSPONGIÆ.
Spongiidæ.
Spongia.
Hercinia.

Aplysinidæ.
Aplysina.
Luffaria = Verongia.
Auliscia.

Chalinidæ.
Chalina.

Phacelliidæ.
Phacellia.

Dysideidæ.
Dysidea.

Halichondriidæ.
Halichondria.

Raphiophora.
Reniera.
Spongilla.

Tethyidæ.
Tethya = Donatia.
Suberites.

Desmacidonidæ.
Isodictya.
Hymeniacidon.
Desmacidon.
Esperia.

Chalinopsidæ.
Axinella.
Chalinopsis.

Geodiidæ.
Pachymatisma.
Geodia.

Ancorinidæ.
Ancorina.

Lithistiidæ.
*Lithistius.
*Corallistes.

Clionidæ.
Cliona.

Xenospongiidæ.
Xenospongia.

Hexactinellidæ.
Euplectella.
Aphrocallistes.
Meyerina.
Hyalonema.
Pheronema.
Asconema.
Dactylocalyx.
Farrea.
Rossella.

———

Dendrospongia.

CALCISPONGIÆ.
Grantiidæ.
Grantia. *arcetta*
Sycon.
Leuconia.
Leucosolenia.
Leucogypsia.

———

*Stromatopora.

Class II. HYDROZOA.

HYDROMEDUSÆ.

Simple or compound organisms, the individual (polypite or hydranth) consisting of a sac composed of an outer (ectoderm) and inner membrane (endoderm), and enclosing a stomach-sac not differentiated from the general body-cavity, the opening furnished with tentacles.

These organisms are nearly all marine ; they are almost invariably soft and gelatinous, occasionally with a chitinous covering (perisarc). Reproduction is either by ova or by zooids, partially independent organisms produced by gemmation or fission ; but sexual communication is requisite at a certain period, or new sex-organs may, after several generations, be developed [alternation of generations]. The reproductive organs are exterior to the body. The digestive cavity communicates directly with the general body-cavity, the outer wall of which is in contact with the water in which the animal lives.

The Hydrozoa are, if we exclude the Rhabdophora, divided into three subclasses :—

Attached to foreign bodies, sometimes in fresh
    water .... HYDROIDA.
Free and oceanic.
    Polypites attached to a disk, float, or body-
    sac .... SIPHONOPHORA.
    A single polypite suspended from the disk... DISCOPHORA.

Subclass I. HYDROIDA.

HYDROPHORA.

Hydrosome fixed, consisting of numerous polypites united together in a branched or tree-like form, and originating from a single polypite, rarely the polypite maintaining a solitary existence. The reproductive elements mostly medusiform.

With a very few exceptions, the polypites are united to form a community or composite organism, which may include 100,000 individuals or personæ.

The character of Hydroida is, according to Dr. Allman, "never with a hydriform strophosome united with the gonosome into a natatory column," which is directly opposed to the character of Siphonophora.

A community of Hydroids has to discharge two dissimilar

functions—alimentation and reproduction; in most cases, at a certain stage, the reproductive element separates, and thenceforth leads an independent existence. The alimentary element is termed the "polypite," and is either single or, more frequently, there is a large number united by the "cœnosarc," which is usually invested by a chitinous covering called the "polypary." In *Sertularida* the polypary is composed of little cells or calycles [hydrothecæ], in each of which an individual is lodged. This is the product of continuous budding.

One of the forms of non-sexual reproduction is the "zooid." Zooids differ from organs in that the zooid is an individual organism, which may or may not be capable of independent existence. A community of zooids in union with one another constitutes the "hydrosome." Zooids are of two kinds: in one, destined for the nutrition of the community, the assemblage is called the "trophosome;" the other gives origin to the generative elements—ova and spermatozoa; and the entire association of these generative zooids is called the "gonosome." The trophosome is composed of the "hydranth" and the "hydrophyton." The hydranth (or polypite) contains the digestive sac; the hydrophyton (or cœnosarc) is, as we have seen, the common basis by which the general community is kept together. The hydrorhiza is the adherent base. The ultimate zooid, which generates either the ova or the spermatozoa, is the "gonophore." The "sporosac" is a gonophore without the umbrella. The "gonozoid" is the sexual zooid, whether fixed or detached and fitted for locomotive life; this is also known as a "medusiform gonophore," or "planoblast." The "gonangium," or "gonotheca," is an external receptacle in which the gonophore is formed.

In the development from the egg the embryo in the ciliated and locomotive stage is known as a "planula." It is, however, not flat, as the name would imply, but conical or cylindrical. In a short time it loses its cilia and, with them, the power of active locomotion, and is gradually changed until it acquires its adult form. The planula consists of ectoderm and endoderm; and the primitive digestive cavity is formed by the invagination of the ectoderm; in *Hydra*, however, the mouth is produced directly from the body-wall. In *Tubularia*, and probably some others, the embryo assumes a radiate appearance, and is then known as an "actinula."

For Claus Hydroida is an order with four suborders—Tabulata, Tubulariæ, Campanulariæ, and Trachymedusæ. For Von Hayek it is a class with six orders—Hydrida, Corynida, Campanularida, Sertularida, Æginida, and Graptolithida.

With a calcareous polypary.
    Polypary, when present, chitinous   ... HYDROCORALLINÆ.
With a hydriform trophosome.
    Not permanently attached .... ELEUTHEROBLASTEA.
    Permanently attached.
        No hydrothecæ nor gonangia......... GYMNOBLASTEA.
        Hydrothecæ and gonangia............ CALYPTOBLASTEA.
Without a hydriform trophosome   ...... HAPLOMORPHA.

Order I. ELEUTHEROBLASTEA.

HYDRIDA. GYMNOCHROA. POECILOMORPHA.

Hydrosome consisting of a single polypite, not permanently fixed. Nutritive buds at maturity discharging themselves and then growing independently as free organisms.

Only one genus is known, containing the common *Hydra viridis* and two or three other species. They are found in fresh water, and if cut up each piece will develop in a few hours into a perfect animal. The body is tubular, capable of great extension; and its proximal end is furnished with a hydrorhiza, by which it can attach itself at will to any foreign body; the opposite end is provided with tentacles, by which it secures its prey.

In the sexual mode of reproduction ova are formed near the fixed end, and spermatozoa, which are liberated at the same time, are formed at the base of the tentacles. *Hydra* is sometimes monœcious, sometimes diœcious.

Hydridæ.
Hydra.

Order II. GYMNOBLASTEA.

CORYNIDA. TUBULARINA. ATHECATA. TUBULARIAN HYDROIDS.

Polypites aggregated. No hydrothecæ nor gonangia present, either for the polypites or the gonophores.

These are delicate plant-like marine organisms, except *Cordylophora*, attached to various foreign bodies, and developing buds [gonophores], which often becoming detached, float away into a free existence [planoblasts, or gonozoids], and are then known under the general name of Medusæ.

The Medusoid gonophore is composed of a swimming-bell (nectocalyx) with its inner margin produced into a delicate membrane called the "velum," its outer margin bearing the tentacles.

From the centre hangs a tubular body—the manubrium—containing the body-cavity, and acting as a polypite. The body-cavity is connected with four or more canals radiating to the circumference, and giving rise with their branches to a circular canal. Nervous filaments with ganglionic enlargements running round the margin have been found in many genera. Pigment-spots (ocelli), black, vermilion, or carmine, are imbedded in the marginal ectoderm. Small corpuscles containing mineral con-cretions, and known as "lithocysts," are supposed to represent auditory organs.

The classification is unusually difficult, owing to the association of similar trophosomes with dissimilar gonosomes, and *vice versâ*. The planoblast also, at its liberation, is in an immature state, and the adult condition is therefore uncertain. In 1864 Allman divided this order into nine families; Carus, the previous year, recognized only two; Hincks, in 1868, had twelve; and in 1871 Allman made twenty-one; eleven of these had only one genus, and mostly only one species in each: they are enumerated below. Schmarda has eight, and Claus ten families, including *Hydridæ*.

Clavidæ.
Clava.
Cordylophora.

Turridæ.
Turris.

Corynidæ.
Coryne.

Syncorynidæ.
Syncoryne.
Gemmaria.

Dicorynidæ.
Dicoryne.

Bimeriidæ.
Bimeria.
Wrightia=Atractylis.
Hydranthea.

Bougainvilliidæ.
Bougainvillia.
Perigonimus.

Eudendriidæ.
Eudendrium.

Hydractiniidæ.
Hydractinia.

Podocorynidæ.
Podocoryne.

[*Cladonemidæ.*
Cladonema.

Nemopsidæ.
Nemopsis.

Pennariidæ.
Pennaria.

Cladocorynidæ.
Cladocoryne.

Myriothelidæ.
Myriothela.

Clavatellidæ.
Clavatella.

Corymorphidæ.
Corymorpha.
Amalthæa.

Monocaulidæ.
Monocaulus.

Tubulariidæ.
Tubularia.

Hybocodonidæ.
Hybocodon.

Hydrolaridæ.
Lar.

Order III. CALYPTOBLASTEA.

SERTULARIDA. CAMPANULARIDA. DIPLOMORPHA. THECAPHORA. SCENOTOCA.

Polypites connected by a cœnosarc, and invested by an unorganized chitinous excretion [polypary or perisarc]. Hydrothecæ and gonangia present.

Plant-like organisms often taken for seaweed, and always attached to some foreign body. The reproductive elements are matured in the gonophore or, in some cases, free medusiform zooids are produced. The capsule or gonotheca is the chitinous receptacle in which the gonophores are formed.

A tubular or cup-shaped extension of the polypary, in which thread-cells [nematophores] are sometimes imbedded, is characteristic of the Plumulariidæ.

In *Ophiodes* Hincks describes certain snake-like organs (cœnosarcal appendages) distributed upon the creeping stolon; "they are vigorous in their movements, capable of enormous elongation, and surmounted by a large capitulum thickly covered with thread-cells." They act as organs of defence and in the capture of food. *Obelia geniculata* is a phosphorescent species, as are also some others of this order.

A new order—Thecomedusæ—has been founded by Allman on a remarkable form, *Stephanoscyphus mirabilis*, found by him at Antibes, imbedded in the substance of a sponge; but whether the association was one of parasitism, or merely accidental, it was impossible to say.

Campanulariidæ.
Clytia.
Campanularia.
Obelia = Laomedea.
Gonothyrea.
Thaumantias.

Campanulinidæ.
Campanulina.

Leptoscyphidæ.
Leptoscyphus.

Lafoëidæ.
Lafoëa.
Salacia.

Trichydridæ.
Trichydra.

Coppiniidæ.
Coppinia.

Haleciidæ.
Halecium.
Ophiodes.

Æquoreidæ.
Æquorea.

Sertulariidæ.
Sertularia.
Diphasia.
Hydrallmania.
Thuiaria.

Plumulariidæ.
Plumularia.
Antennularia.
Aglaophenia.

Order IV. HAPLOMORPHA.

CRYPTOCARPÆ. CRASPEDOTA. HYDROPHORA. MONOPSEA. TRACHYMEDUSÆ.

No hydriform trophosome; medusæ developing directly from the ovum.

These are the true Medusæ; but at present the order can only be regarded as provisional. The embryo, so far as is known, is directly developed from the parent without passing through any intermediate form as in the medusiform gonophore. Both seem able to produce independent forms like themselves by budding. Many naked-eyed Medusæ (Gymnophthalmata of Forbes) are now known to be the free gonophores of the Gymnoblastea and the Calyptoblastea. Some are very minute, others attain the size of a walnut.

The Acalephæ of Cuvier comprised this order, Discophora, except *Lucernaria*, Siphonophora, and Ctenophora.

Trachynemidæ.	Cunina.	Liriope.
Trachynema = Circe.	Ægineta.	Carmarina.
Rhopalonema.	Polyxenia.	
Aglaura.		*Charybdæidæ.*
	Geryoniidæ.	Charybdæa.
Æginidæ.	Geryonia.	
Ægina.		

Order V. HYDROCORALLINÆ.

Polypary calcareous, sometimes divided into compartments by transverse partitions. Mouth of the polypites with or without tentacles.

The cœnosarc is made up of a "network of anastomosing canals," and its outer layer is provided with thread-cells. The zooids are of two kinds; the smaller and more numerous have no mouth and no stomach-cells. Reproduction is by means of gonophores.

The Milleporidæ, "in the general form of their zooids, seem allied to the gymnoblastic hydroids, whereas the presence of distinct gonangia in the Stylasteridæ seems to ally the latter to the calyptoblastic group" (*Moseley*).

Milleporidæ.	Stylaster.
Millepora.	Errina.
	Allopora.
Stylasteridæ.	Polypora.
Cryptohelia.	

Distichopora.

Subclass II. DISCOPHORA.

PHANEROCARPA.   MEDUSÆ.   ACRASPEDA.

Hydrosome consisting of a single disk, from which one or more polypites are suspended.

Free-swimming oceanic animals, whose reproduction is some-times by buds, which are formed either in pouch-like organs, dilatations of the stomach, or from the tentacles, or from the sides of the polypite. They are the Steganophthalmata or covered-eyed Medusæ of Forbes; and to them Claus has restricted the old Cuvierian name of Acalephæ. They are well known as sea-blubbers and sea-jellies. Most of them are luminous, but they do not appear to possess any special light-giving organs.

Schmarda still (1877) includes in the Discophora the naked-eyed Medusæ of Forbes, which are now known to be the sexual zooids or gonophores of the Gymnoblastea.

There are three orders: but Calycozoa are sometimes ranked as a subclass; by Schmarda they are placed with the Anthozoa.

Capable of attachment by the proximal end... CALYCOZOA.
Incapable of attachment.
    Polypites numerous........................... RHIZOSTOMEA.
    Polypite single .... MONOSTOMEA.

Order I. RHIZOSTOMEA.

Polypites numerous, modified with the genitalia into a root-shaped mass. No central mouth nor marginal tentacles.

At the extremity of the arms of the root-like mass are small openings, through which the food is conveyed along a central canal to the stomach.

The embryo is a free oblong body [planula] which, soon attaching itself to some foreign substance, forms a mouth and stomach by invagination; tentacles then arise from the mouth, in which stage it is known as the "hydra tuba" (Scyphistoma), which by budding gives rise to colonies of "Hydriform polypi;" some of these assuming the form of a pile of cups placed one within the other, is now called a "strobila;" then the cups separating, each becomes a free-swimming disk [ephyra], by degrees acquiring the adult form.

Rhizostomidæ.	*Cassiopeiidæ.*	*Polycloniidæ.*
Rhizostoma.	Cassiopeia.	Polyclonia.
Leptobrachiidæ.	*Cepheidæ.*	*Crambessidæ.*
Leptobrachia.	Cephea.	Crambessa.

*Hexarhizites.

Order II. MONOSTOMEA.

PELAGIADA.

Polypite single. A central mouth. The disk with marginal tentacles, or if without them with tentacles under the disk.

The reproductive elements are developed in the disk, or by fission from a fixed trophosome, and, in their detached condition, grow with great rapidity, ultimately attaining a weight of many hundreds of pounds. *Cyanæa arctica* has been found with a disk seven feet in diameter and with tentacles fifty feet long, the fixed trophosome from which it proceeded being of very small size.

Pelagiidæ.	*Cyaneidæ.*
Pelagia.	Cyanea.
Chrysaora.	
Nausithoë.	*Aureliidæ.*
	Aurelia = Medusa.

Order III. CALYCOZOA.

PODACTINARIA.   LUCERNARIIDA.

Polypite single, in the centre of a cup-shaped umbrella, its proximal end fixed. Generative elements discharging themselves into the body-cavity.

The umbrella is eight- or nine-lobed in *Lucernaria*, each lobe bearing a tuft of tentacles; in *Carduella* they form one continuous series. "The whole organism is semitransparent, variously coloured, and of a gelatinous consistence."

To the form of disk without a velum Huxley restricts the term umbrella: in the Calycozoa it is prolonged aborally into a longer or short peduncle, terminating in a hydrorhiza, by which the animal is enabled to fix itself to any foreign body at will. When detached, the contractions of the umbrella enable it to swim with the ease of an ordinary medusoid body. The order contains but one family:—

Lucernariidæ.
Lucernaria.
Depastrum.
Carduella.

Subclass III. SIPHONOPHORA.

OCEANIC HYDROZOA.

Hydrosome free and oceanic, simple or branched, consisting of several polypites connected by a contractile cœnosarc, and attached at the proximal end to a disk, float [pneumatophore], or body-sac [somatocyst].

The polypites have each a tentacle, often of great length, provided with lateral branches having thread-cells [trichocysts, modified zooids]. Each polypite is occasionally protected at the base by overhanging plates [hydrophyllia]. Certain bell-shaped cups [specialized nectocalyces] are frequently present, attached to the hydrosome, by the contraction of which the animal is propelled through the water. The pneumatophore contains an air-sac [pneumatocyst], variously formed, with firm chitinous walls. Vesicles and pigment-spots [ocelli], often very brilliant, are found round the margins of the nectocalyces: the former have been called "otolites," and have been supposed to be auditory organs; the latter are possibly the earliest indication of eyes.

The Siphonophora are organisms of a very delicate and peculiarly composite character, almost exclusively found floating on the surface of tropical seas. They have rarely a radiate character, but are either bilateral or unsymmetrical. Their reproductive organs are gonophores, varying from the simplest form to medusoids of the normal type.

A body-sac at the proximal end......... CALYCOPHORÆ.
A float at the proximal end .... PHYSOPHORÆ.

Order I. CALYCOPHORÆ.

Polypites united by a filiform and unbranched cœnosarc; the proximal end modified into a somatocyst, and provided with one or more nectocalyces.

"Sets of appendages—each consisting of a hydrophyllium, a hydranth with its tentacle, and gonophores, which last bud from the pedicle of the hydranth—are developed at regular intervals on the cœnosarc, and the long chain trails behind as the animal swims with a darting motion, caused by simultaneous rhythmical contraction of its nectocalyces, through the water." [*Huxley.*] The distal set of these appendages, as they attain their full development, "becomes detached as a free-swimming complex *Diphyzooid.*" In this condition they grow and alter their form, until the gonophores which they develop "become detached, increase in size, become modified in form, and are set free as a third series of independent zooids."

These animals are so transparent as only to be noticed at a distance by their bright tints.

Diphyidæ.	*Sphæronectidæ.*	*Hippopodiidæ.*
Diphyes.	Sphæronectes.	Hippopodius.
Abyla.	Monophyes.	Vogtia.

Prayidæ.

Praya.

Order II. PHYSOPHORÆ.

Polypites united by an unbranched, or very slightly branched, filiform, globular, or discoidal cœnosarc; the proximal end modified into a pneumatophore, and sometimes provided with nectocalyces. Mostly monœcious.

The tentacles are either attached to the cœnosarc, or singly to a polypite; they are forty inches long in *Halistemma rubrum,* while the pneumatophore is only three or four lines in its largest diameter. The pneumatophore, however, is generally of much larger size, and in the Velellidæ it is "converted into a sort of hard inner shell, its cavity being subdivided by septa into numerous chambers."

The members of this order differ considerably among themselves, but they all agree in having a pneumatophore.

The well-known "Portuguese man-of-war" (*Physalia pelagica*) is the only species of the order that has received an English name. It represents a suborder (order) for Claus, as do also the Velellidæ (Discoideæ).

Apolemiidæ.	*Physophoridæ.*	*Physaliidæ.*
Apolemia.	Physophora.	Physalia.
	Stephanospira.	
Stephanomiidæ.		*Velellidæ.*
Stephanomia.	*Athorybiidæ.*	Velella.
Agalma.	Athorybia.	Porpita.
Halistemma.		
	Rhizophysidæ.	
	Rhizophysa.	

Class III. ACTINOZOA.

POLYPI. CORALLARIA. ANTHOZOA.

The digestive cavity not in contact with the outer wall of the body, but separated by an intervening perivisceral space.

The perivisceral space is radially divided into a number of compartments by membranous partitions [mesenteries], in which the reproductive organs are placed. Reproduction also takes place by budding, by fission of small fragments separating from the edge at the base [*Gosse*], as well as by ordinary generation. The egg, in the latter case, develops into an infusorial-like germ, with vibratile cilia and free locomotion [planula]. The sexes are either united or distinct.

The mouth is furnished with tentacles, hollow, and either simple or fringed, capable of being withdrawn into the body-cavity. No manducatory apparatus exists.

The great majority are composite organisms, mostly provided with a horny or calcareous secretion, known as the "corallum" or "polypary." The corallum-tissue [sclerenchyma] presents every gradation between the solid condition and the spicular stage. For these animals collectively Huxley extends the term "Coralligena."

There are two subclasses:—

Tentacles simple .... ZOANTHARIA.
Tentacles pinnately fringed .... ALCYONARIA.

Subclass I. ZOANTHARIA.

POLYACTINIA.

Polypes with simple or occasionally branched tentacles, six, or a multiple of six. Corallum, when present, mostly sclerodermic, more rarely sclerobasic.

The corallum is a hard, mostly calcareous substance [calcium carbonate], secreted externally in the sclerobasic corals, and internally in the sclerodermatous; in the Malacodermata it is, when present, disseminated in the form of small spicules [sclerites]. An individual [persona] of the compound corallum is known as a "corallite;" the outer wall forms the "theca," the upper part of which is the cup or calicle; below it is often divided radially by distinct vertical septa [mesenteries] united in the centre to the "columella." Sometimes the thecæ are divided by horizontal plates [tabulæ]. "Dissepiments" are incomplete plates growing from the sides of the septa. The "cœnenchyma" is the calcareous covering uniting the individual corallites together; it is secreted by the cœnosarc, with which it may be said to correspond.

Milne-Edwards divides the Zoautharia into three groups:—

Corallum either absent or rudimentary .... MALACODERMATA.
Corallum present
    Corallum internal, calcareous .... SCLERODERMATA.
    Corallum external, not calcareous .... SCLEROBASICA.

Order I. MALACODERMATA. (Sea-anemones.)

HELIANTHOIDA. ACTINIARIA. HEXACORALLA.

Corallum absent, or represented by a few spicules. Tentacles numerous, simple. Polypes rarely aggregated. Sexes mostly distinct.

In the Zoanthidæ only are the polypes united by a common creeping stem. The majority, owing to their muscular base, enjoy some power of locomotion, although generally adherent to some foreign body. The Ilyanthidæ have no adherent base; and *Arachnactis* is a free-swimming organism: it is, however, possibly an immature form.

The tentacles are generally disposed in two or more series. These are successively developed from within outwards, in multiples of six; but one or more tentacles are sometimes abortive. They are often perforate at the end. In *Sagartia bellis*, a common species, there are six rows, the inner minute, and altogether they amount to about 500. [*Gosse.*]

Zoanthidæ.	*Actiniidæ.*	Anthea = Anemonia.
Zoanthus.	Thalassianthus.	Actinia.
Palythoa.	Adamsia.	
	Tealia.	*Ilyanthidæ.*
Cerianthidæ.	Bolocera.	Edwardsia.
Cerianthus.	Bunodes = Cereus.	Ilyanthus.
	Actinoloba.	Peachia.
Minyadidæ.	Sagartia.	Halcampa.
Minyas.	Capnea.	———
Nautactis.	Corynactis.	Arachnactis.
	Aiptasia.	

Order II. SCLEROBASICA.

ANTIPATHARIA.

Corallum external, not calcareous. Tent ˙les six, simple. Polypes united, included within the corallum.

There is a rough, stem-like, branched, horny axis, or cœnosarc, covered by a very friable cœnenchyma, which generally becomes detached in drying.

Few species of this order are known, and they are mostly from the Mediterranean. One of the species of *Antipathes* has a tapering cœnosarc nine feet long, with the diameter at the base ·3 of an inch.

| *Antipathidæ.* | Arachnopathes. | *Gerardiidæ.* |
| Antipathes. | Leiopathes. | Gerardia. |

Order III. SCLERODERMATA. (Stone-corals.)

MADREPORARIA. LITHOCORALLIA. CORALLIGENA.

Corallum internal, calcareous.   Tentacles more than six. Polypes generally united.

The corallum-tissue is firm and compact in the Eporosa, porous and granular, or even spongy, in the Perforata. The Rugosa (Tetracoralla) are only known from the remains of extinct forms; in these there are four septa, in all the others six, at least in the young state.

The coral-reefs of warm seas are built up by the members of this order, mostly Astræidæ. The largest of these, the Great Barrier Reef, is 1200 miles long, and 20 to 100 broad.   The coral does not grow below a depth of 25 or 30 fathoms, and not in water under a temperature of 66°.

There are five suborders:—

Corallum hexameral.
Septa rudimentary or absent.
Tabulæ well developed .... TABULATA.
No tabulœ............................... TUBULOSA.
Septa present.
Corallum porous ....,.... PERFORATA.
Corallum imperforate.................. EPOROSA.
Corallum tetrameral [Palæozoic] .... RUGOSA.

The Tabulata and the Rugosa have been placed by Carus among the Hydrozoa, forming his group (?) Lithydrodea.

RUGOSA.

Stauriidæ.

*Stauria.
*Holocystis.

Cyathaxoniidæ.

*Cyathaxonia.
*Guynia.

Cyathophyllidæ.

*Zaphrentis.
*Cyathophyllum.
*Strombodes.
*Lonsdaleia.

Cystiphyllidæ.

*Cystiphyllum.

TABULATA.

Seriatoporidæ.

Seriatopora.

Favositidæ.

*Favosites.
Pocillopora.

Theciidæ.

*Thecia.

TUBULOSA.

Auloporidæ.

*Aulopora.
*Pyrgia.

PERFORATA.

Madreporidæ.

Turbinaria.
Madrepora.
Manopora.
Eupsamma.
Balanophyllia.
Dendrophyllia.

Poritidæ.

Porites.
Montipora.
Alveopora.
Psammocora.

EPOROSA.

Turbinoliidæ.

Turbinolia.
Caryophyllia.

Dasmiidæ.

Dasmia.

Oculinidæ.

Oculina.
Amphihelia.

Stylophoridæ.

Stylophora.
Madracis.

Astræidæ.

Euphyllia.
Favia.
Meandrina.
Astrangium.
Astræa.

Echinoporidæ.

Echinopora.

Merulinidæ.

Merulina.

Fungiidæ.

Fungia.
Agaricia.
Ctenactis.

D

Subclass II. ALCYONARIA.

Asteroida. Octactinia. Octocoralla.

Polypes with eight pinnately fringed tentacles in one series. Corallum, when present, external, spicular, or with a sclerobasic axis, but occasionally thecal or tubular.

The polypes are connected by the cœnosarc, through which permeates prolongations of the body-cavity of each, thus permitting a free circulation of fluids. There is sometimes an outer skeleton, either with or without a central sclerobasic axis. The corallum is rarely thecal, "never presenting traces of septa."

These composite organisms are, with few exceptions, fixed; they are only found in deep water.

Adherent to some foreign body.
 Ectoderm leathery, contractile .... ALCYONIACEÆ.
 Ectoderm hard, not contractile.
  Branched.
   Calcareous or horny .... GORGONIACEÆ.
   Alternately calcareous or horny ... ISIDACEÆ.
  Tubular...................................... TUBIPORACEÆ.
Not adherent ....:.............. PENNATULACEÆ.

Order I. ALCYONIACEÆ.

Ectoderm leathery, slightly contractile, with calcareous spicules in tissues. No sclerobasis. Permanently rooted.

Alcyonium digitatum, a lobed, spongy-looking body, pellucid when distended with water, and covered with stellate apertures for the polypes, is well known under the name of "Dead men's fingers." *Telesto* is a tree-like organism, with a tubular, subcalcareous corallum.

Alcyoniidæ.	*Cornulariidæ.*
Alcyonium=Lobularia.	Cornularia.
Nephthya.	Anthelia.
Sarcophyton.	Sarcodictyon.

Telestidæ.
Telesto.

Order II. GORGONIACEÆ.

Axis branched, erect, sclerobasic, either horny or calcareous : permanently rooted. Cœnenchyma smooth.

The axis is sulcate, by which it is known, *inter alia*, from Antipathidæ. The branches are frequently anastomosing. In Briareidæ there is no horny axis, but the internal parts are composed of calcareous spicules. *Corallium* has a sclerobasic homogeneous stony axis, and should probably form an order of its own. Its only representative is the red coral of commerce, *Corallium rubrum.*

" *Heliopora* seems to differ from all other Alcyonarians except *Corallium.*" [*Moseley.*]

Primnoidæ.	Gorgonia.	*Gorgonellidæ.*
Primnoa.	Leptogorgia.	Gorgonella.
Muricea.	Plexaura.	
	Eunicea.	*Briaridæ.*
Gorgoniidæ.	Siphonogorgia.	Briareum.
Rhipidigorgia.		Paragorgia.
		Semperina.

Coralliidæ.	*Helioporidæ.*
Corallium.	Heliopora.

Order III. ISIDACEÆ.

Axis branched, erect, composed of alternate calcareous nodes and horny internodes; permanently rooted. Polypes embedded in the cœnosarc.

The nodes are larger than the internodes and are generally white and fluted. In *Melithæa* the nodes are porous or corky in appearance.

Isididæ.	*Melithæidæ.*
Isis.	Melithæa.
Mopsea.	

Order IV. TUBIPORACEÆ.

Corallum sclerodermic, in the form of tubular thecæ, bound together by horizontal plates [epithecæ]; no septa. Polypes completely retractile.

The horizontal plates are confined to the outside of the corallum, which is perforated by numerous minute canals. The " organ-pipe coral " (*Tubipora musica*), forming large hemispherical masses, is typical of this order, of which only one genus is known; its polypes are violet or grass-green in colour.

Tubiporidæ.

Tubipora.

Order V. PENNATULACEÆ.

Polypary free, the basal end without polypes, the upper end generally variously branched and bearing the polypes.

The polypes or zooids are mostly of two kinds, one set being sexually developed, the other set sexless. These are borne on the upper part of a fleshy cœnosarc, provided with a horny flexible internal axis.

These organisms mostly live with one end stuck deeply in the mud; but the Pennatulidæ are free, moving, however, languidly, and do not swim. Among the few British species one, known as the "sea-pen" (*Pennatula phosphorea*), is highly phosphorescent, and so probably are the entire group; its ova are carried at the back of the pinnæ.

Umbellularia has a rod-like axis six feet in length, with a tuft of polypes at its apex. It is quite an aberrant form.

Pennatulidæ.

Pennatula.
Pteroeides.
Sarcoptilus.

Virgulariidæ.

Virgularia.

Scytalium.
Pavonaria.

Veretillidæ.

Veretillum.
Lituaria.

———

Umbellularia.

Renillidæ.

Renilla.

Cophobelemnidæ.

Cophobelemnon.

Class IV. CTENOPHORA.

CILIOGRADA.

Gelatinous, transparent organisms, swimming by means of rows of cilia, mostly disposed in comb-like plates [ctenophores]. No corallum. Hermaphrodite.

The ctenophores consist of eight meridional bands, lying between the two poles marking the opposite extremities of the body, and dividing the interpolar region into an equal number of

lobes [actinomeres] ; each band has a number of successive ridges
or plates, to each of which a row of cilia is attached.   The lateral
actinomeres contain each a sac, which gives rise to the tentacles
when present.

In this class occurs a well-marked sense-organ [ctenocyst], ovate
or spherical, occupying a central position, which " would seem to
be the localized recipient of those obscure general impressions to
which its lowly-organized possessor is capable of responding."
[*Greene.*]

Reproduction is by ova, which are expelled by the mouth,
and the young are gradually developed, few having a larval stage.
The homologies between this class and the larval echinoderms
have led to the suggestion of their being united.

The species vary considerably in size and shape.   They are all
oceanic and of a very delicate texture, leaving only a mere film
when dried in the sun ; they are said, however, to be very voracious,
and to have a powerful digestion.

With two oral lobes............................ LOBATÆ.
Without oral lobes.
    Body ribbon-shaped....................... TÆNIATÆ.
    Body oval or round.
      With two filiform tentacles............ SACCATÆ.
      Without tentacles .... EURYSTOMATA.

Order I. LOBATÆ.

Body with two oral lobes.   Tentacles various, turned towards
the mouth or wanting.

The lobes project from the antero-posterior regions of the body,
sometimes so as to conceal the mouth.   The two lateral actinomeres generally terminate in a slender appendage or tentacle.
Eurhamphæidæ have no tentacles.

According to Agassiz, *Sicyosoma* is the larval form of a *Cestum.*

Eurhamphæidæ.	*Mnemiidæ.*	*Calymmidæ.*
Eurhamphæa.	Mnemia.	Calymma.
	Lesueuria.	
Bolinidæ.	Chiajea.	*Ocyroidæ.*
Bolina.		Ocyroë.

Sicyosoma.

Order II. TÆNIATÆ.

Body ribbon-shaped, without oral lobes. Tentacles two, near the mouth.

Each half of the ctenophoral system is represented by four very long canals.

The common Mediterranean species, known as " Venus's girdle " (*Cestum veneris*), has a flat, ribbon-shaped body, three or four feet in length, and a height of about two inches.

Cestidæ.

Cestum.

Order III. SACCATÆ.

Body oval or spheroidal, without oral lobes. Tentacles two, away from the mouth.

The filiform tentacles are fringed in *Cydippe* on one side, and can be withdrawn instantly into the sac of the lateral actinomeres, at the will of the animal.

This and the two preceding orders are united by Von Hayek under the name of Stenosomata.

Cydippidæ.
Cydippe = Pleurobrachia.
Eschscholtzia.

Gegenbauria.
Owenia.

Callianiridæ.
Callianira.

Mertensiidæ.
Mertensia.

Order IV. EURYSTOMATA.

Body oval or oblong, without oral lobes. No tentacles. Mouth very large.

The mouth and digestive cavity are large, unlike the remainder of the class. *Beroë* itself is of the size and shape of a lemon.

Beroidæ.

Beroë.
Idyia.
Pandora.

Neisidæ.
Neis.

Rangiidæ.
Rangia.

The following group is made a subclass of Hydrozoa by Nicholson, a suborder of Hydroida by Allman, while Carus (1875) places it with the Polyzoa, as one of the four "orders" of Gymnolæmata. It is Von Hayek's sixth order of Hydroida; but is placed as a pendant to the Alcyonaria by Schmarda. It is not noticed by Claus.

RHABDOPHORA.

GRAPTOLITHINA.

"Hydrosoma compound, occasionally branched, consisting of numerous polypites united by a cœnosarc, the latter being included in a strong, tubular polypary, whilst the former were protected by hydrothecæ." (*Nicholson.*)

These are Palæozoic organisms, generally known as "Graptolites," whose structure is still far from being entirely understood. They "are usually found as glistening, pyritous impressions, with a silvery lustre. In some cases, however, they are found in relief." The genera are numerous.

*Monograpsus.
*Diplograpsus.
*Rhabdopleura.

Subkingdom III. ECHINODERMATA.

Marine animals, widely different in appearance, but with more or less of a radiate structure. An alimentary canal distinct from the body-cavity, and mostly with an anal aperture. A vascular system in most. Deuterostomatous. Sexes mostly distinct.

The egg generally develops a free-swimming, ovoid, ciliated, bilateral larva or pseudo-embryo, called a "*Pluteus*." A process of modification or secondary development within the embryo or larva, and absorbing its digestive organs, leads to the adult form; the parts of the larva therefore do not correspond with those of the adult, they are either absorbed or cast off. There is, however, often a direct development, the young being reared within or upon the body of the parent, and retaining a commensal relation with her until able to provide for themselves.

All Echinoderms have a calcareous skeleton, and many are provided with movable spines. A characteristic apparatus of vessels, termed the ambulacral or water-vascular system, is present. It is composed of a ring round the pharynx, from which proceed a number of radiating canals, commonly giving off cæcal appendages [Polian vesicles] as well as branches, which enter the retractile tube-feet, often furnished with a terminal disk or sucker [ambulacra], which, with the spines, are the organs of locomotion. The "madreporic canal" connects the pharyngeal ring with the exterior. "The ambulacral vessels are filled with a fluid containing numerous nucleated cells."

Originally described as parasitic animals are the "pedicellariæ" (homologous with the spines, according to A. Agassiz), found in some members of the Echinoidea and Asteroidea. They are small slender bodies having a soft skin, with two or three claws at the top, which they open and shut with great activity; their use is supposed to be for cleansing purposes.

The Echinodermata are said to have no annectant forms. The Gephyrea resemble certain Holothurioidea; but the structural difference between their larvæ are such that they "can never be genealogically connected." A. Agassiz, however, considers that there is "nothing in nature to justify their separation" as a subkingdom from Cœlenterata (1877).

There are four classes; but the Crinoidea are sometimes included in the Stellerida.

Body stalked .... CRINOIDEA.
Body not stalked.
    An external shell of calcareous plates. ECHINOIDEA.

No shell.
> Body lobed or stellate .... STELLERIDA.
> Body elongated or vermiform ...... HOLOTHURIOIDEA.

Class I. CRINOIDEA.

PALMATOZOA.

Mostly star-shaped animals, fixed during the whole or part of life by a stalk or column.

The body or calyx of the ventral surface is directed upwards; the stalk is on the dorsal or inferior portion.

The greater part of the Crinoidea are extinct. Eight genera are known in a recent state.

There are three orders; perhaps *Edriaster* should form another order, as has been done by Huxley. It is not supposed to have had any stalk.

With arms.
> Body rounded .... CYSTOIDEA.
> Body cup-shaped .... CRINOIDEA.
> Without arms .... BLASTOIDEA.

Order I. CYSTOIDEA.

Body rounded; enclosed in numerous pentagonal, suturally united plates, and furnished with a jointed stalk; at the side "a large aperture, closed by a pyramid formed of triangular plates" (*Huxley*). Palæozoic.

The use of the "pyramid" is unknown; some suppose it to be the "oro-anal" orifice. *Hyponome*, said to belong to this order, has been recently found in Torres Straits.

The principal genera are:—

*Caryocrinus.	*Codaster.	*Comarocystites.
*Hemicosmites.	*Apiocystites.	*Sphæronites.

*Edriaster.

Order II. CRINOIDEA.

BRACHIATA.

Body cup-shaped, the dorsal portion furnished with calcareous plates, the ventral coriaceous; stalked, at least when young; and with five or more branching arms, not connected with the visceral cavity.

This order abounded in Palæozoic times. *Comatula*, the commonest of the recent forms, is free when mature. The mouth is on the ventral surface; the arms constantly in motion cause a sufficient quantity of organic matter in solution to pass into it for food. Some of the extinct forms had a proboscis or tube, often of great length, arising between two of the arms.

There are two families, the first permanently stalked, the other free in the adult. J. Müller, however, divides the Crinoidea into "Articulata" and "Tessellata," the latter comprising *Platycrinus*, *Actinocrinus*, and some others, all extinct.

Encrinidæ.	Hyocrinus.	*Comatulidæ* (Feather-
Pentacrinus (Sea-lily).	*Encrinus.	stars).
Holopus.	*Apiocrinus.	Comatula=Antedon.
Ophiocrinus.	*Platycrinus.	Phanogenia.
Rhizocrinus.	*Actinocrinus.	Actinometra.
Bathycrinus.		*Saccosoma.

Order III. BLASTOIDEA.

Body rounded, enclosed in solid polygonal plates, and furnished with a jointed stalk; the summit of the body, or calyx, with five areas longitudinally grooved. No pyramid. Palæozoic.

There are about fifty species; the majority appear to have lived in the Carboniferous period.

*Pentatremites.   *Elæacrinus.   *Eleutherocrinus.

Class II. STELLERIDA.

ASTEROIDEA.

Body star-shaped or pentagonal, composed of a disk, either five or more lobed, or with five or more arms. Mouth central, inferior, without teeth. Sexes distinct.

The integument is coriaceous, strengthened by a vast number of calcareous plates. One or more madreporiform tubercles or plates are connected with the ambulacral water-system. The anal aperture is either dorsal or absent. A blood-vascular system is present in most.

Fossil species are numerous, extending from the Lower Silurian *Asteracanthion*, which still exists, up to the present time. *Goniaster* and *Astropecten* are found in the London Clay.

Disk entire, with five or more arms ...... OPHIUROIDEA.
Disk lobed or pentagonal.................... ASTEROIDEA.

Order I. OPHIUROIDEA.

Disk entire, containing all the viscera, and giving off five or more slender arms, not channelled beneath. Tube-feet without suckers.

The arms are not prolongations of the disk, and are not channelled beneath, as in Asteroidea; they are each enclosed in four rows of calcareous plates, and, having no suckers, these arms are their only means of locomotion. There are no pedicellariæ. The madreporiform tubercle is inferior, near the mouth.

J. Müller has two divisions—Ophiureæ with simple, and Euryaleæ with branched arms; the latter is confined to the family Astrophytidæ. The former are known as "sand-stars." Our commoner British species are *Ophioglypha lacertosa, Ophiura lævis, Ophiothrix rosula*, and *Amphiura squamata* [= *Ophiocoma neglecta, Johnst.*].

Ophiodermatidæ.
Pectinura.
Ophioderma=
   Ophiura.
Ophioglypha.

   Ophiolepididæ.
Ophiolepis.

   Amphiuridæ.
Ophiopholis.

Ophiactes.
Amphiura.
Hemipholis.

   Ophiocomidæ.
Ophiocoma.
Ophiomastix.

   Ophiothricidæ.
Ophiothrix.
Ophiopsila.

Ophiomyxidæ.
Ophiomyxa.
Ophioscolex.

   Astrophytidæ.
Astrophyton=
   Euryale.
Asteronyx.
Trichaster.
Asteroschema.

Astrophiuridæ.
Astrophiura.

Order II. ASTEROIDEA. (Starfishes.)

Disk more or less lobed or pentagonal, the lobes continuous with the disk, hollow, and receiving prolongations of the viscera. Tube-feet with suckers.

The lobes are channelled beneath for the ambulacral feet or suckers, which act as organs of locomotion. The madreporiform tubercle is on the back; a curved calcareous canal connects it with the mouth. Sometimes five tubercles are present, and as many canals. Pedicellariæ occur on the body or on the mouth.

Brachina of Van Beneden is supposed to be a larval condition of *Asteracanthion rubens* (Fivefinger of the oyster-dredgers), our common species. *Bipinnaria* and *Brachiolaria* are among the successive larval stages of the species of this order.

Asteracanthiidæ.	Scytaster.	*Asterinidæ.*
Asteracanthion = Uraster.	*Palæaster.	Asterina.
	Oreaster.	Asteriscus = Palmipes.
Heliaster.	*Asteriidæ.*	Culcita.
	Asterias = Astropecten.	Goniaster.
Solastridæ.	Ctenodiscus.	*Brisingidæ.*
Echinaster = Cribella.	Luidia.	Brisinga.
Chætaster.	*Palæocoma.	
Solaster.		

Class III. ECHINOIDEA.

Body rounded or discoidal, enclosed in a shell composed of numerous closely connected calcareous plates, and studded with tubercles on which are jointed movable spines. Sexes distinct.

Certain of the plates are perforated for the emission of the tube-feet. These are the "ambulacral" plates, arranged alternately in pairs; between each pair are the "interambulacral" plates, also in pairs; usually there are five pairs of each. The anus is at the top, surrounded by two series of small plates, the inner known as the "genital," the outer as the "ocular" plates; each of the latter is perforated and "supports the eye-spot." The mouth is below, and furnished with a single series of "oral" plates. The latter has usually a very complicated arrangement of jaws or teeth [the splanchnic skeleton]. The intestine is convoluted. There is a blood-vascular system, but no distinct organs of respiration.

Sea-water is admitted into the peritoneal cavity and into the water-vascular system through the perforated madreporiform tubercle, which is borne on the largest of the five genital plates.

Except in Cidaridæ, there are found in the ambulacral areas certain minute hyaline ciliated bodies called "sphæridia;" Lovén supposes them to be sense-organs.

In the Pluteus-stage of the larva there is an internal keleton of calcareous rods as in the Ophiuroidea.

The first two orders of the following table form the Euechinoidea of Bronn; the last order is known as "Tessellata," and is now extinct.

Two rows of ambulacral alternating with two
   rows of interambulacral plates.
    Anus central...... .... ENDOCYCLICA.
    Anus not central .... EXOCYCLICA.
Three to six rows of plates in each ambulacral
   space .... PALÆCHINOIDEA.

Order I. ENDOCYCLICA.

DESMOSTICHA. REGULARIA.

Mouth and anus central. Two rows of ambulacral and two
rows of interambulacral plates alternating with each other.

The plates overlap one another in some of the extinct forms;
in *Asthenosoma* and *Phormosoma* they are movably connected by
membrane.

In our common sea-urchin (*Echinus sphæra*) there are about
300 plates and 4000 spines (*Forbes*). The Mediterranean *E.
esculentus* is extensively used for food; the ovaries are the parts
selected. *Toxopneustes lividus* is found in holes of limestone,
sandstone, granite, &c., which it is supposed to form for itself.

Cidaridæ.	*Echinothuriidæ.*	*Echinidæ.*
Cidaris.	*Echinothuria.	Toxopneustes.
*Goniocidaris.	Phormosoma.	Heliocidaris.
Salenia.	Asthenosoma=Cal-	Hipponoë.
Arbaciidæ.	veria.	Echinus (Sea-urchin).
Arbacia.	*Echinometridæ.*	Salmacis.
Diadematidæ.	Echinometra.	Mespilia.
Diadema.	Podophora.	Echinothrix.

Order II. EXOCYCLICA.

PETALOSTICHA. IRREGULARIA.

Anus not central. Two rows of ambulacral and two rows of
interambulacral plates alternating with each other.

The masticatory apparatus is frequently absent. The mouth
is central in Clypeastridæ and Mellitidæ, but excentric in the remaining families. The two former are placed by A. Agassiz
among the Endocyclica, by Claus they are ranked as a separate
order (Clypeastridea).

The English species of this order are known as "heart-
urchins." As they mostly bury themselves in the sand or mud,
the greater part of their spines are directed backwards.

Clypeastridæ.
Echinocyamus.
Fibularia.
Clypeaster.
Laganum.

Mellitidæ.
Mellita.
Rotula.
Encope.
Echinarachnius.
*Scutella.

Cassidulidæ.
*Cassidulus.
Echinoneus.
Echinolampus.
*Ceratomus.
*Dysaster.
_____

Anochanus.

Spatangidæ.
Ananchytes.
Pourtalesia.

Spatangus.
Agassizia.
Breynia.
Palæostoma=Leskia.
Amphidetus = Echinocardium.
Brissus.
Metalia.
Meoma.
Linthia=Desoria.
Schizaster.
Moira.

Order III. PALÆCHINOIDEA.

TESSELLATA.

Three to six rows of plates in each interambulacral space.
This order is confined to the Palæozoic formations. There is but one family :—

Palæchinidæ.
*Palæchinus.
*Melonites.
*Eocidaris.

Class IV. HOLOTHURIOIDEA.

SCYTODERMATA.

Body cylindrical or vermiform, with a coriaceous skin, in which are scattered calcareous particles, and, with few exceptions, furnished with five longitudinal rows of ambulacral suckers or tube-feet. Mouth tentaculate.

The tentacles surrounding the mouth are plumose and retractile, and generally from ten to twenty in number; they are prolongations of the water-vascular system. The anal aperture is at the opposite extremity. The intestine is convoluted and often complicated. The skin is lined with powerful longitudinal and transverse muscles, by which the animal is enabled so to contract its body as to eject all its viscera. In this condition it will live for some time, and in three or four months the lost parts will be reproduced. Some members of this class will divide spontaneously into two or more parts, each developing new and perfect organs.

Notwithstanding, the anatomy, especially of the water-vascular system, is very complex.

Reproduction is sometimes direct, but a bilateral planula, it is said, generally emerges from the egg. *Synapta* has a form of larva known as "Auricularia."

The "Cuvierian organs" are thread-like tubes and fibres opening into the cloaca, or into the water-vascular system.

No fossil species are known; a few of their calcareous spicules have been found in the Carboniferous deposits.

Claus has two orders—Pedata and Apoda: the first with two families—Aspidochirotæ (*Holothuria*, &c.) and Dendrochirotæ (*Psolus*, &c.); the second order contains the two families Molpadidæ (constituting his suborder Pneumonophora) and Synaptidæ.

Without organs of respiration .... APNEUMONA.
With organs of respiration .... PNEUMONOPHORA.

Order I. APNEUMONA.

No special organ of respiration. Hermaphrodite. No Cuvierian organs.

There are no ambulacral feet in Synaptidæ; but locomotion is effected by the spicules, which are anchor-shaped, barbed, wheel-shaped, &c., according to the genera. In Oncinolabidæ the tube-feet are in five rows and the tentacles are filiform.

Synaptidæ.	Chirodota.	*Oncinolabidæ.*
Synapta.	Eupyrgus.	Oncinolabes.
Myriotrochus.	Anapta.	Echinosoma.

Order II. PNEUMONOPHORA.

DIPNEUMONA.

Respiratory organs branching, tree-like, opening into the cloaca. Sexes distinct. Cuvierian organs present.

Water is admitted into the abdominal cavity by means of the respiratory organs : these are two in number, except in *Rhopalodina*. Molpadiidæ have no tube-feet ; they are doubtfully hermaphrodite. The tentacles are either shield-shaped [Aspidochirotæ = Holothuriidæ] or branched [Dendrochirotæ = Psolidæ]. In Molpadiidæ the tentacles are either cylindrical or shield-shaped. *Rhopalodina* has a flask-shaped body, the mouth at the smaller end surrounded by tentacles; it has four branchiæ. It forms the order Tetrapneumona of Schmarda [=Diplostomidea, Semper; Decacrenidia, Bronn].

Some *Holothuriæ* harbour inquilines or commensals (fish, mollusks, crustaceans). *Holothuria argus,* a large black species, is the trepang, or *bêche de mer,* an article of considerable importance as food in the Eastern seas.

Molpadiidæ.

Molpadia.
Liosoma.

Holothuriidæ.

Holothuria (Sea-cucumber).
Sporadipus.

Psolidæ.

Psolus.
Cucumaria=
    Pentacta.
Thyone.
Ocnus.

Thyonidium.
Phyllophorus.
Hemicrepis.

———

Rhopalodinidæ.

Rhopalodina.

Subkingdom IV. **VERMES.** (Worms.)

Body generally elongate or vermiform, soft, bilaterally symmetrical, with or without feet, but, when present, never jointed.

There is a great diversity of form and structure in this subkingdom, so that little can be said of it collectively. There is a
water-vascular system, and, occasionally, "pseudohæmal vessels
are present. A digestive cavity is sometimes absent. A few have
eyes, but sense-organs are of the simplest kind. Reproduction is
mostly by ova. Many species are external (Ectozoa) or internal
parasites (Entozoa).

The water-vascular system is a tubular set of vessels having
openings on the surface of the body, and branching out into its
substance. It is never subservient to locomotion, as in the Echinodermata.

The classes composing this subkingdom are widely different
from one another, and were for a long time combined with
Arthropoda, under the name of Annulosa or of Articulata.
There are many points of approximation between them and the
Mollusca.

A division of the classes has been made into Archæostomata
(mouth the same through life) and Deuterostomata (mouth in
the adult a secondary development). The former includes Platyelmintha except Cestoda, Nematelmintha, Gephyrea, and Rotifera; the latter Annelida except Hirudinea, Chætognatha, and
Polyzoa. Giard unites Annelida, Polyzoa, and "satellite groups"
to the Mollusca, constituting his Gynotoca; while Nematoda,
Gastrotricha, &c. are his Nematelmia.

Normally single animals.
    Tail never fin-like.
        Body distinctly segmented.................. ANNELIDA.
        Body not segmented, or very indistinctly.
            No ciliated disk.
                Mouth simple or none.
                    Body flat........................... PLATYELMINTHA.
                    Body rounded...................... NEMATELMINTHA.
                Mouth at the end of a proboscis ... GEPHYREA.
            A ciliated disk at the anterior end ... ROTIFERA.
    Tail fin-like............................. CHÆTOGNATHA.
Compound plant-like organisms .... POLYZOA.

E

Class I. PLATYELMINTHA.

STERELMINTHA. SCOLECIMORPHA. COTYLIDEA. PLATODES.

Body flat, more or less elongated, without true segments. Mouth sometimes wanting. Hermaphrodite.

These worms are of a low organization. They have, with few exceptions, no circulatory or respiratory systems, and their nervous system consists at most of two pharyngeal ganglia giving off few branches. Some of the species are rather round than flat.

The Scolecida of Huxley comprised that portion of his Annuloida other than Echinodermata. In the Cotylidea of Van Beneden, Schmarda includes Cestoda, Trematoda, and Hirudinea.

For Giard this class alone comprises the Vermes, and with them he places *Dicyema* (class Dicyemida) and his genera *Rhopalura* and *Intoshia* (class Orthonectida), the orders below being also ranked as classes. The Orthonectida are minute, ciliated animals, parasitic on Ophiuridæ and Lineidæ. Dicyemida are parasites apparently confined to Cephalopoda. "The renal organs of most *Sepiæ* may be said to be literally made up of these organisms in all stages of growth" (*Ray Lankester*). According to Van Beneden they are pluricellular animals, forming by themselves one of the principal divisions of the animal kingdom, which he has named Mesozoa.

> With a digestive cavity.
>   Cuticle ciliated; free .... TURBELLARIA.
>   Cuticle not ciliated; parasitic........... TREMATODA.
> With a digestive cavity .... CESTODA.

Order I. TURBELLARIA.

PLATYELMINTHA. TERETULARIA.

Non-parasitic, aquatic animals, having a flattened, ovoid, or elongate body, with a ciliated cuticle. Mostly hermaphrodite.

Except in the Nemertidæ the sexes are rarely distinct. Reproduction is either by ova, by internal gemmation, or by transverse fission. The intestine is either simple or branched, and in some there is no circulatory system. There are eye-specks in most. Some Nemertidæ begin life as a helmet-shaped larva [pilidium]. A peculiar proboscidiform modification of the pharynx has been taken for a genital organ, for an embryo, for the digestive canal, for an organ of defence, and for a parasitic worm.

A few species are found in damp earth or in fresh water, others are inquilines in Holothurioidea or in the respiratory cavities of Ascidians.

Ehrenberg, M°Intosh, and others, rank the Turbellaria as a class; and it is variously divided. The families below are distributed under three sections :—Dendrocœla with a ramified digestive cavity, Rhabdocœla with a simple one, both being without an anal aperture (Aprocta), and Rhynchocœla with an anal aperture (Proctucha) and the pharynx (proboscis) protrusible and furnished with stylets. The first two sections also form the suborder Planarida of some systematists, distinguished by their oval or elliptic form; the third section, constituting the Nemertoidea, have an elongated linear form, one—*Lineus marinus* (the sea long-worm)—sometimes attaining a length of 200 feet [*M°Intosh*]. Of the Planariidæ some are very minute. Geoplanidæ comprise the "land-planarians."

Pelagonemertes is a transparent leaf-shaped animal with a dendrocœle intestine. *Pterosoma* appears to be an allied form.

Balanoglossus, forming the class Enteropneusti of Gegenbauer, has a complex branchial apparatus. *Tornaria* is supposed to be its larval form. *Mitraria* is also supposed to be a larval form belonging to this order. Convolutidæ constitute the order Accœla of Ulianin, and is adopted by Schmarda.

Dendrocœla are divided by Stimpson into Monogonopora (single sexual aperture), including Planariidæ and Geoplanidæ; and Digonopora (double sexual aperture), including Euryleptidæ, which has been split up into many families. Rhynchocœla are also divided into Enopla (proboscis armed with stylets) = Amphiporidæ, and Anopla (proboscis unarmed), including the remaining families. These have mostly received character-names. In this order there is very little accord between Schmarda and Claus as to the position of the genera or to the sequence of the families. In the Rhynchocœla, M°Intosh's classification has been adopted.

RHABDOCŒLA.	Derostomum.	*Convolutidæ.*
	Mesopharynx.	Convoluta.
Microstomidæ.	Opisthomum.	Nadina.
Microstomum.		
Dinophilus.	*Mesostomidæ.*	*Rhynchoprobolidæ.*
	Mesostomum.	Rhynchoprobolus.
Derostomidæ.	Macrostomum.	Prostomum.
Vortex.		Alaurina.

Catenulidæ.

Catenula.

DENDROCŒLA.

Planariidæ.

Polycelis.
Planaria.

Geoplanidæ.

Bipalium.
Geoplana.
Polycladus.

Euryleptidæ.

Stylochus.

Planocera.
Leptoplana.
Thysanozoon.
Cephalolepta.
Eurylepta.

RHYNCHOCŒLA.

Cephalothricidæ.

Cephalothrix.
Ommatoplea.

Carinellidæ.

Carinella.
Valencinia=Polia.

Balanoglossus.

Lineidæ.

Lineus.
Borlasia.
Meckelia.
Micrura.

Amphiporidæ.

Nemertes.
Tetrastemma.
Amphiporus.

———

Pelagonemertes.
Pterosoma.

Order II. TREMATODA.

External or internal parasites, flattened or rounded, not ciliated in the adult state, and provided with one, two, or more ventral suckers. Mouth and anus in one. Mostly hermaphrodite.

The alimentary canal lies in the substance of the body, and not in a free perivisceral space; in a few it disappears in the adult, and is sometimes much branched.

Some of these parasites scarcely undergo any change [Monogenea]; others [Digenea] begin life on leaving the egg as a free ciliated infusorian; if it meets with a suitable host in its wandering, it puts on many forms before reaching maturity. Individuals proceeding from ciliated embryos also produce buds which develop numerous tadpole-shaped larvæ [cercariæ], which, when their ciliated skin has been thrown off, are known as "rediæ." Cercariæ often become encysted; and in that state they are said to wait for years before their host is swallowed by the creature intended to lodge them. The cyst is then broken up, and the worm is set free to begin another form of life.

These parasites, of which there are about 500 species, are found on the gills and skin of fishes, on mollusks, crustacea, &c., and in the eyes, blood-vessels, and intestines of man and other animals. *Distoma hepaticum* is the liver-fluke of sheep. In *Bilharzia hæmatobius*, common in the Egyptians, the slender female is lodged in an abdominal groove of the larger and stouter male.

Myzostoma, an anomalous form, constituting the suborder Ecto-parasita of Von Hayek, is found on *Comatulæ*.

MONOGENEA.		DIGENEA.
Octocotylidæ.	Ancyrocephalus. Octobothrium. Polystomum. Tetrastoma.	*Distomidæ.*
Octocotyle. Diplozoon. Choricotyle. Anthocotyle.		Distoma (Fluke). Bilharzia.
Gyrodactylidæ.	*Tristomidæ.*	*Monostomidæ.*
Gyrodactylus. Calceostoma.	Tristoma=Nitzschia. Udonella. Epibdella.	Monostomum. Amphilina.
Polystomidæ.		*Amphistomidæ.*
Onchocotyle.		Amphistomum. Diplostomum. Gastrodiscus.
	Myzostomidæ. Myzostoma.	

Order III. CESTODA.

TÆNIADA. AGASTREÆ.

Internal parasites, mostly of an elongated and flattened form ; the anterior end of the body or head provided with hooks or spines, or both, or suckers. No digestive nor vascular systems. Hermaphrodite.

In the more typical forms these parasites are composed of a head, which is the true animal, the joints being the hermaphrodite reproductive organs developed by a process of gemmation from the head. The joints are called proglottides, and are organically connected by the water-vascular system. There is only one proglottis in Caryophyllæidæ ; but in *Tænia solium* there are sometimes as many as 800.

Although the proglottides are only produced in the alimentary canal of man or some other warm-blooded animal, it is necessary for the evolution of an embryo that the ovum should be swallowed by some other animal than the one inhabited by the mature worm. When the fecundated proglottides, therefore, are expelled, the decomposing ova are liberated ; and should an ovum get into the alimentary canal of a vertebrate, the embryo (now called a "proscolex"), set free from its covering, proceeds to bore with its spines through the tissues of its host until it finds a resting-place ;

then it surrounds itself with a cyst, and a vesicle containing a fluid is developed; it is now called a "scolex." These cysts were also known as "hydatids."

"When ova of the pork tape-worm (*Tænia solium*) gains access to the alimentary canal of a pig, their shells become digested, and the enclosed six-hooked embryos escape and bore their way into the circulation. Thence they proceed to the cellular tissue and become transformed into measles (*Cysticercus cellulosæ*). In the sheep the cystic worm of the brain (*Cænurus cerebralis*), which causes the 'gid' or 'staggers,' becomes the *Tænia cænurus* of the dog. The *Cysticercus pisiformis*, or pea-measle of the rabbit, is the scolex of *Tænia serrata* infesting the dog. The *Cysticercus fasciolaris* of the mouse becomes the *Tænia crassicollis* of the cat. The common hydatid (*Echinococcus veterinorum*) becomes the *Tænia echinococcus* of the dog. The slender-necked hydatid (*Cysticercus tenuicollis*) of the sheep becomes the *Tænia marginata* of the dog. The *Cysticercus talpæ* and *C. longicollis* infesting moles become respectively the *Tænia tenuicollis* and *T. crassiceps* of the fox. Lastly, a scolex called *Staphylocystis micracanthus*, which is found in a myriopod (*Glomeris*), is the larval state of *Tænia pistillum* infesting shrews (*Sorex*)." [*Dr. Cobbold in lit.*]

The chain of reproductive joints or zooids is called the "strobila." Each new joint is formed between the head and the next joint; the most distant joints are therefore the oldest.

The Cestoda are sometimes combined with the Trematoda and the Turbellaria to form the class Platyelmintha; or the two former only are combined to form the class Cotylidea.

According to Cobbold, there are about 260 species belonging to the Cestoda.

Tæniidæ.	*Diphyllidæ.*	*Tetrarhynchidæ.*
Tænia (Tape-worm).	Echinobothrium.	Tetrarhynchus.

Dibothriidæ.	*Tetraphyllidæ.*	*Caryophyllæidæ.*
Dibothrium.	Phyllobothrium.	Caryophyllæus.
Bothriocephalus.	Acanthobothrium.	Eustemma.

Ligulidæ.
Ligula.

Class II. NEMATELMINTHA.

Body more or less cylindrical or thread-shaped, without true segments, and without limbs. Mouth anterior, often furnished with hooks or papillæ.

These worms are mostly entozoic; they have generally a body-cavity and a distinct nervous system. Eye-specks or ocelli are frequently present. The water-vascular system assumes many forms, but is sometimes wanting. Development is mostly direct.

Schmarda includes Chætognatha in this class. About 1400 species are known. They are divided into two orders :—

With a body-cavity .... NEMATODA.
Without a body-cavity .... ACANTHOCEPHALA.

Order I. NEMATODA.

CŒLELMINTHA. NEMATELMINTHA.

Mostly internal parasites, with thread-like or cylindrical non-ciliated bodies. No respiratory nor circulatory organs. Sexes distinct.

These have a distinct mouth and intestine, although in some Gordiidæ they are either rudimentary or wanting. Anguillulidæ and Enoplidæ have no nervous system. A few are subject to metamorphosis; thus a *Urolabes*-form is the larva of *Filaria*, and a *Rhabditis*-form of *Dochmius*. *Ascaris*, and probably some others, are dimorphous. The young cast their skins before arriving at maturity.

The Gordiidæ are sometimes ranked as an order; *Gordius* itself leads at first a free existence in water, but soon finds its way into some aquatic insect. Its species are sexless so long as they are parasitic. *Mermis* is found in Lepidoptera, and *Sphærularia bombi* in the humble bee. The female of the latter, formerly mistaken for the male, has the uterus 28,000 times larger than herself; this uterus was supposed to be the female. This excessive development of the uterus in some of these worms causes the obliteration of the ordinary opening; and the young only escape by the bursting of the maternal body.

Anguillulæ are found in stale paste, vinegar, ears of wheat affected with blight, &c.; while some of them cause galls on plants. The too well-known guinea-worm (*Filaria medinensis*) varies from six inches to ten or twelve feet in length, but is scarcely two thirds of a line in thickness. *Syngamus trachealis* is found in the trachea of birds, and is the cause of the " gapes." Among the parasites in man may be mentioned *Ascaris lumbricoides* and *Trichina spiralis*, the latter getting in millions into the muscles, and causing the disease called trichinosis or trichiniasis ;

its original home is said to be the rat. *Dochmius duodenalis* is another parasite from which one fourth of the population of Egypt suffers more or less, and often fatally. The great kidney-worm of dogs and wolves, *Strongylus gigas*, attains a length of three feet. It has been also met with in man.

Chætosoma (the Rhabdophora of Claparède), in some respects resembling *Sagitta*, is, with *Cystoopsis* and *Desmoscolex*, obscure forms, placed as pendants to this order by Schmarda.

The following list of families and genera is from Claus. He, however, has not mentioned certain genera which for Diesing and, after him, Schmarda are the types of families, *i. e. Hedruris*, *Cheiracanthus*, and *Ophiostoma* (=Dacnitidæ). For the two families of free Nematoids—Anguillulidæ and Enoplidæ (many species of both are marine)—Bastian adopts thirty genera in his monograph of the Anguillulidæ.

Ascaridæ.
Oxyuris.
Ascaris.
Heteracis.

Strongylidæ.
Strongylus.
Syngamus=Sclero-stomum.
Dochmius.
Ollulanus.
Physaloptera.
Cucullanus.

Trichinidæ.
Trichina.
Trichocephalus.

Filariidæ.
Filaria.
Spiroptera.
Ancyracanthus.

Mcrmididæ.
Mermis.
Sphærularia.

Gordiidæ.
Gordius.

Anguillulidæ.
Anguillula.
Tylenchus.
Rhabditis.

Enoplidæ.
Dorylæmus=Uro-labes.
Enchelidium.
Enoplus.

Chætosomidæ.
Rhabdogaster.
Chætosoma.

Order II. ACANTHOCEPHALA.

Internal parasites, worm-like, the anterior end armed with re-curved hooks. No mouth nor alimentary canal. Sexes distinct.

The water-vascular system is said to have no external openings. The embryos are like Gregarinida, and become encysted as in Cestoda ; in that state they are swallowed by birds, fishes, &c., when they develop into mature animals. There are about 100 species.

Echinorhynchidæ.
Echinorhynchus.
Coleops.

Class III. CHÆTOGNATHA.

ŒSTHELMINTHA.

Free, elongate, transparent animals, with preoral setæ and a fin-like tail, but without limbs. Hermaphrodite. No metamorphosis.

The body is fringed on each side by a striated fin-like membrane. The intestine is a straight tube. The nervous system consists chiefly of a single abdominal ganglion, sending backwards and forwards two pairs of lateral branches. The eyes are two pigment-spots. The embryo is non-ciliated, and its development is, in some respects, peculiar.

This is a very aberrant form, and has even been regarded as a vertebrate. Forbes placed it among the Mollusca. There are six species, varying from half to one inch in length, mostly European, of the one genus

Sagitta.

Class IV. GEPHYREA.

Sipunculacea. Rhyngodea.

Body cylindrical, with a thick coriaceous skin, often indistinctly ringed. No limbs. Head not distinct from the body, often produced into a proboscis. Sexes distinct.

There are no respiratory organs, and sometimes no vascular system. The mouth, with or without tentacula, is placed at the end of a retractile proboscis. The anus is either at the base of the proboscis, or is at the posterior extremity of the body. Eye-specks are present in certain Sipunculidæ. A *Planaria*-like form of male occurs in *Bonellia*. These animals have ciliated free-swimming embryos, not unlike Rotifera, to which Huxley considers them closely allied. In the subsequent growth only part of the larva is concerned in the development of the adult (an alternation of generation, or metagenesis).

The Gephyrea, of which there are about 120 known species, vary from half to eight or more inches in length. They live among rocks or seaweed, or are sometimes found buried in the sand.

Formerly classed with the Echinodermata, they are now divided into two, somewhat artificial, orders—Inermes (without bristles and without a double blood-vascular system), including Sipunculidæ and Priapulidæ; and Armata or Chætiferi (with corneous bristles and a double blood-vascular system), = Echiuridæ and Sternaspidæ. The latter is, however, referred by Claus to Polychæta. *Chætoderma*, placed in Priapulidæ by the same authority, has the body furnished with spines.

Sipunculidæ.	*Priapulidæ.*	Bonellia.
Sipunculus=Syrinx.	Priapulus.	Thalassema.
Phascolosoma.	Anoplosomatum.	Ancistropus.
Aspidosiphon.	Chætoderma.	
Diesinga.		*Sternaspidæ.*
Dendrostomum.	*Echiuridæ.*	Sternaspis.
	Echiurus.	

Class V. ANNELIDA.

ANNULATA.

Body composed of numerous segments [somites], the limbs rarely present, or, if present, rudimentary [parapodia]. A vascular system with red blood, but without corpuscles, in most.

The skin is soft, composed of many layers, the surface mostly ciliated, the cilia bristle-like, and often fasciculate; the segments are sometimes to the number of 400. There is occasionally a rudimentary head [prostomium]; it bears two or more cirri or tentacles (antennæ and palpi), sometimes, but rarely, jointed. The blood-vascular system is very variable, but is composed mainly of longitudinal canals. The blood is rarely propelled from behind forwards.

There are no true parasites in this class, except some Hirudinea; but inquilines or commensals are sometimes met with. Very few species are terrestrial.

The genera *Polyophthalmus*, *Dero*, and *Capitella* form the Haloscolecina of Carus. Anarthropoda is a divisional name, used to include this and the two preceding classes.

Mouth suctorial .... HIRUDINEA.
Mouth not suctorial.
    Without branchiæ .... OLIGOCHÆTA.
    With branchiæ.
        Branchiæ dorsal .... CHÆTOPODA.
        Branchiæ cephalic .... CEPHALOBRANCHIA.

Order I. HIRUDINEA. (Leeches.)

DISCOPHORA. SUCTORIA.

Mostly aquatic animals, with a sucking-disk at one or both extremities. Hermaphrodite. No metamorphosis.

Although ringed, the body is not divided into distinct somites, the rings being merely surface-markings. Respiration is effected by the skin or by lateral sacs. The circulatory system consists mostly of longitudinal trunks or a series of sinuses. The eyes are little more than pigment-spots. The skin is without cilia.

Leeches are mostly freshwater animals, swimming easily ; but many are parasites on fishes, Mollusca, and Crustacea. A few are terrestrial, occurring in Japan, Ceylon, Chili, &c.

The ordinary leech is *Hirudo medicinalis* ; the Hungarian leech (*H. officinalis*) is probably a variety. Horse-leech is a name given to two distinct species, or even genera—*Hæmopsis sanguinea* and *Aulostomum gulo*. *Macrobdella valdiviana* is 2½ feet long.

Malacobdellidæ.
Malacobdella.

Acanthobdellidæ.
Acanthobdella.
Histriobdella.

Branchiobdellidæ.
Branchiobdella.

Clepsinidæ.
Clepsine.
Piscicola = Ichthyo-bdella.

Hirudinidæ.
Typhlobdella.
Macrobdella.
Bdella.

Trochetia.
Aulostomum.
Nephelis.
Hirudo (Leech).
Hæmopsis.
Pontobdella (Sea-leech).
Branchellion.

Order II. OLIGOCHÆTA.

LUMBRICINA. TERRICOLA. SCOLEINA. ABRANCHIATA.

Mostly land- or freshwater-worms, without feet, but provided with bristles [setæ]. Mouth rudimentary. No branchiæ. Mostly hermaphrodite. No metamorphosis.

Reproduction is mostly by ova or by gemmation ; but the earth-worm (*Lumbricus terrestris*) develops directly. Respiration is by the skin.

These worms are generally found in the mud of ponds and streams ; they are never parasitic or inquiline. They have great powers in repairing injuries.

Ichthydium (worm-shaped, with a ciliated ventral surface) is by some referred to the Rotifera; Metschnikoff forms for it an order—Gasterotricha. Schmarda, the more recent authority, places it in this order, and he also includes in it "Polyophthalmida," Maldanidæ (= Clymenidæ), and Chætopteridæ. With the two following orders they form his sixteenth Class—Chætopoda (*Van Beneden*). He estimates that these three orders contain 1500 species.

Ichthydiidæ.	*Enchytræidæ.*	*Lumbricidæ.*
Ichthydium.	Enchytræus.	Lumbricus (Earth-
Chætonotus.	Chætogaster.	worm).
	Dero = Proto.	Perichæta.
Naididæ.		
Æolosoma.	*Sænuridæ.*	*Phreoryctidæ.*
Nais = Stylaria.	Sænuris = Tubifex.	Phreoryctes.
Aulophorus.	Euaxes = Rhynchelmis.	Megascolex.
	Limnodrilus.	

Order III. CHÆTOPODA.

ERRANTIA. ANTENNATA. POLYCHÆTA. RAPACIA. NOTOBRANCHIATA.

Marine animals, mostly worm-like, with tubular setigerous feet [parapodia]. Body not presenting distinct regions. Metamorphosis in most.

The sexes are mostly distinct, but asexual forms occur giving rise to the former; "a process of zooid development" takes place which "appears to be a combination of fission with gemmation;" and "the result is, not infrequently, the formation of long chains of connected zooids." The males and females, in some cases, differ from one another as well as from the sexless forms. The young, on leaving the ovum, is a free-swimming ciliated body. *Syllis, Eunice,* and a few others are said to be viviparous.

The branchiæ are external and ranged along the side of the back; but in many the branchiæ are rudimentary or absent, and the respiration is either carried on by the skin or by sea-water admitted to the perivisceral cavity: mixed with chyle, this fluid is known as the "chylaqueous fluid." According to Owen it serves the place of an internal skeleton by acting as the base of resistance to the cutaneous muscles, "the power of voluntary motion being lost when the fluid is let out." The blood-vascular system is sometimes in abeyance. The blood is red in most, but in *Aphrodite* it is yellow, in a few it is colourless. It is contained in two long tubes, one dorsal, the other ventral. The presence of corpuscles is disputed.

The head is generally provided with two or four tentacles. The mouth contains one or more pairs of jaws (or teeth); and the pharynx forms a protrusible proboscis. Eyes are sometimes present. In *Alciope* there is a retina, iris, and lens.

A few species are commensals; but they are mostly free, moving about at the bottom of the sea, or living in the sand, as the common lob-worm (*Arenicola piscatorum*). Apparently they are capable of living a long time without food; McIntosh records a *Eunice norvegica* kept for three years without nourishment of any kind. *Eunice gigantea* attains a length of four feet. Several species are phosphorescent. "Tracks and burrows of Annelids are found commonly in rocks of all ages from the Cambrian period upwards."

Tomopterus, forming the order Gymnocopa of Grube, is a degraded form, with no trace of branchiæ or of a blood-system. Although three or four inches long, it is so transparent as to be nearly invisible, except in certain shades of light. In the female the ova are in the general cavity of the body, there being no proper ovarium. *Polygordius* is a transitional form towards Trematoda: Schmarda places it as a pendant to Chætopoda. Of its two species, *P. purpureus* is hermaphrodite, while *P. luteus* is diœcious.

The following list of families and genera is compiled principally after Schmarda, except that *Polyophthalmus*, Clymenidæ, and Chætopteridæ are placed by him in Oligochæta: these families are placed by Claus in this order, which, for him, also includes Sedentaria; he has, however, referred the first nine families in the list below to the same group (Cephalobranchia).

Capitellidæ.
Capitella=Lumbriconais.
Notomastus.
Dasybranchus.

Opheliidæ.
Ophelia.
Ammotrypane.
Polyophthalmus.

Arenocolidæ.
Arenicola (Lobworm).
Eumenia.

Clymenidæ.
Clymene.
Maldane.
Ammochares = Owenia.

Ariciidæ.
Aricia.
Ephesia.
Aonis.

Cirratulidæ.
Cirratulus.
Audouinia.
Cirrhinereis.

Nerinidæ.
Nerine = Malaco-
　ceros.
Spio.
Disoma.

Leucodoridæ.
Leucodora.

Chætopteridæ.
Chætopterus.

- *Syllidæ.*
Syllis = Autolytus.
Grubea.
Dujardinia.
Schmardia.

Hesionidæ.
Hesione.
Castalia.
Psamathe.
Podarce.

Phyllodocidæ.
Phyllodoce.
Eulalia.
Alciope.

Nephthyidæ.
Nephthys.

Glyceridæ.
Glycera.

Nereidæ.
Nereis = Lycoris.
Lycastis.

Lumbriconereidæ.
Lumbriconereis.
Lysidice = Palolo.
Œnone.

Eunicidæ.
Eunice = Marphysa.
Diopatra.

Amphinomidæ.
Hipponoë.
Chloëia.
Euphrosyne.
Amphinome.

Aphroditidæ.
Aphrodite (Sea-
　mouse).
Hermione.
Pontogenia.
Harmothoë.
Polynoë.
Pholoë.
Antinoë.
Sigalion.
Pelogenia.

Palmyridæ.
Palmyra.

———

Tomopteridæ.
Tomopterus.

Order IV. CEPHALOBRANCHIA.

SEDENTARIA. TUBICOLÆ. POLYCHÆTA.

Worm-like, marine animals, mostly protected by a tube. Body presenting distinct regions. Respiration by branchiæ placed near or on the head. Sexes distinct. A metamorphosis.

The tubes are mostly secreted by the animal itself, and are either calcareous or membranous, or they may be composed of grains of sand agglutinated together. The tubes are either free or adherent to some foreign body, and the animal is not organically attached to them. The branchiæ are generally filamentous and fringed with vibrating cilia, but they are sometimes absent. The blood in *Sabella* is olive-green. There is no proboscis. The embryo is a free-swimming ciliated body.

A few fossil forms are known. *Spirorbis* and other tubicolar annelids occur as early as the Silurian period. Some forms, as *Ditrupa*, have been taken for shells of mollusks allied to *Dentalium.*

Pherusidæ are free, but are often adherent ten or a dozen together to one *Echinus*. *Phoronis* is placed in Gephyrea by Claus. In some of the Serpulidæ the carbonate of lime is secreted in such masses that the small circular reefs formed everywhere round the Bermudas are due to their agency alone.

The following families alone compose this order, according to Schmarda:—

Pherusidæ.
Pherusa = Chloræma.
Siphonostoma.
Flemingia.

Hermellidæ.
Hermella = Sabellaria.
Pallasia.

Terebellidæ.
Terebella.
Amphitrite.

Polycirrus.
Sabellides.

Amphictenidæ.
Amphictene = Pectinaria.

Sabellidæ.
Sabella.
Myxicola.
Branchiomma.
Amphiglena.

Serpulidæ.
Serpula.
Vermilia.
Cymospira.
Protula = Apomatus.
Ditrupa.
Spirorbis.
Filograna.
Fabricia.
Phoronis.

Class VI. ROTIFERA.

ROTATORIA. SYSTOLIDA.

Minute, aquatic, rarely parasitic animals, mostly free-swimming. The head generally provided with one or more ciliated disks. Sexes distinct. No metamorphosis.

The body is cylindrical, more or less distinctly ringed. At the anterior end is one or more retractile disks [trochal disk or corona], bearing cilia, which, when vibrating rapidly, produce the appearance of turning round like a wheel: hence they are sometimes called "wheel-animalcules." The mouth is ventral. The pharynx contains the biting and grinding machinery, and is known as the "mastax." Certain red spots, sometimes to the number of eight, are generally supposed to be eyes. The males are frequently smaller than the females, and have no digestive organs. So far as they are known they are shortlived; "they seem to be simply locomotive organs of fecundation, whose services are occasionally required;" the females carrying on the process of reproduction alone. The winter ova differ from the summer ova in being enclosed in a peculiar shell.

Some genera are provided with a tube or sheath [lorica], into which the animal can withdraw itself. *Philodina, Melicerta,* and one or two others construct a floccose covering for their protection in winter.

The Rotifers delight in the sunshine; on a dull day they conceal themselves. They swim either by means of their cilia, or by bending the body and jerking themselves along. There are about 200 species, nearly all freshwater. Flosculariidæ are permanently fixed. *Notommata parasita* is an entozoon of *Volvox globator.*

By some these animals are regarded as unsegmented Arthropods. Ehrenberg places Ichthydiidæ in this class, under his order Monotrocha. *Albertia* is worm-shaped, and parasitic on *Lumbricus, Nais,* &c. *Seison* is found on *Nebalia;* these two genera constitute the Perosotrocha of Schmarda, a pendant to the Rotifers. *Trochosphæra,* with a globular body and a ciliated girdle, is another pendant. *Apsilus* and *Balatro* are destitute of vibratory cilia; they are present, however, in the young of the former.

No intestine nor anus .... GASTRODELA.
With intestine and anus.
  One disk.
    Disk entire.................................. HOLOTROCHA.
    Disk divided .... SCHIZOTROCHA.
  Two disks .... ZYGOTROCHA.

GASTRODELA.

Asplanchnidæ.

Asplanchna.
Ascomorpha.

HOLOTROCHA.

Ptyguridæ.

Ptygura.
Diplotrocha.

Œcistidæ.

Œcistes.
Conochilus.

SCHIZOTROCHA.

Megalotrochidæ.

Megalotrocha.

Flosculariidæ.

Floscularia.
Melicerta.
Stephanoceros.

Hydatinidæ.

Hydatina.
Diglena.
Eosphora.
Notommata.
Synchæta.
Polyarthra.
Apsilus.

Euchlanidæ.

Euchlanis.
Stephanops.
Metopidia.

ZYGOTROCHA.

Philodinidæ.

Philodina.
Rotifer.
Actinurus.

Scaridiidæ.

Scaridium.

Brachionidæ.

Brachionus.
Anuræa.
Noteus.

Albertiidæ.

Albertia.
Seison = Saccobdella.
Balatro.

Class VII. POLYZOA.

BRYOZOA. CILIOBRANCHIATA. POLYPIARIA. CELLULIFERA.
TENTACULIFERA. TENTACULIBRANCHILE.

Associated, mostly marine animals, each living in a cell on a plant-like organism, or adnate upon foreign bodies, very rarely free. Mouth surrounded with ciliated tentacles. Hermaphrodite. A metamorphosis.

The separate animals or zooids are called "polypides" or "personæ," the entire system is a "zoarium," also "polyzoarium" or "cœnœcium," and the "zoœcium," or cell, is the organ in which the polypide is contained. The tentacles are borne on a projection or stage called the "lophophore." The alimentary canal "is suspended in a double-walled sac." There is a single nerve-ganglion placed between the mouth and the anus, which are very near together. There is no vascular system, nor any sense-organ. The animals do not communicate with one another as they do in the hydroid polypes [Calyptoblastea]. There is, however, said to be a general system of nerves by which the polypides are placed in communication.

Reproduction is by ova, the young appearing in the form of a free ciliated sac-like body ; and by continuous gemmation, each new zooid remaining attached, and adding to, the parent-stock. Another form of gemmation, confined, however, to the Phylactolæmata, occurs in which certain bodies, called "statoblasts," are developed in the polypide ; these are enclosed in two horny disks, and "when the statoblasts are placed under circumstances favouring their development, they open by the separation from one another of the two faces, and then there escapes from them a young *Polyzoon*, already in an advanced stage of development, and in all essential respects resembling the adult individual in whose cell the statoblasts were produced" (*Allman*).

Much difference of opinion exists as to the affinities of the Polyzoa. Ehlers considers that their nearest allies are the Gephyrea ; Claus places them between the Nematoid worms and the Rotifera ; Barrois also insists on their intimate relations to Rotifers. Von Hayek (1877) unites the two orders Bryozoa and Rotatoria to form his Ciliata, the sixth and last class of Vermes. They have also been held to constitute a lower form of Tunicata ; but then, as has been observed, no Polyzoon begins life as an Ascidian, and no Ascidian begins life as a ciliated gemmule ; and if, as Hancock

F

has suggested, the branchial sac of the Ascidian is (as are the endostyle, tentacular filaments, &c.) a new and distinct development, then they have no homological representations in the Polyzoa.

Recently Ray Lankester places this class with Brachiopoda and Lamellibranchiata to form his Lipocephala; Huxley associates the Polyzoa with the Brachiopoda as "a division apart," for which he proposes the term "Malacoscolices." Brooks maintains that the union of the Brachiopoda and Polyzoa is "without scientific value." Lastly, Schmarda (1878) keeps the old name "Molluscoidea" for this class and Tunicata.

Cyphonautes is a remarkable form not yet determined; it has been said to be the larva of *Membranipora*.

Schmarda gives 600 living and 1800 fossil species for this class.

Mouth without an epistome........ .... GYMNOLÆMATA.
Mouth with an epistome .... PHYLACTOLÆMATA.

Order I. GYMNOLÆMATA.

STELMATOPODA. INFUNDIBULATA. ECTOPROCTA.

Separation of individuals complete. No epistome. Lophophore circular.

This order is composed of marine organisms, often taken for seaweeds, and having a close resemblance to the Calyptoblastea; the former, however, differ in that the cells do not communicate except by the external investing cuticle [ectocyst]. In one of the suborders [Chilostomata] this cuticle gives rise to certain appendages called "avicularia," "vibracula," and "oœcia," or "oocysts." The avicularia are organs of prehension, consisting of a movable portion or mandible and a corresponding fixed portion; the vibracula consist of a cup having a movable seta attached to it; the oœcia or oocysts are globular cells receiving the ova.

The sequence of the families in the following list is after Busk; a later arrangement by Claus omits many of the families and genera of the former, while, on the other hand, the latter gives two or three families not noticed by Busk.

There are three suborders :—

Mouth of the cells with a movable lip ...... CHILOSTOMATA.
Mouth of the cells without a lip.
    Mouth round, open, not setose .... CYCLOSTOMATA.
    Mouth setose................................... CTENOSTOMATA.

CHILOSTOMATA.—Cell-opening or mouth closed with a movable lip or operculum. Avicularia and vibracula mostly present.

Claus divides this suborder into four groups:—*a. Cellularina*, zoœcium corneous, funnel-shaped: Æteidæ, Cellulariidæ, and Bicellariidæ. *b. Flustrina*, zoœcium quadrate, smooth externally: Flustridæ, Cellariidæ, and Membraniporidæ. *c. Escharina*, zoœcium mostly calcareous, quadrate or semioval, with a lateral opening: Eschariporidæ, Myriozoidæ, Escharidæ, and Discoporidæ. *d. Celleporina*, zoœcium calcareous, rhomboid or oval, with a terminal mouth: Celleporidæ and Reteporidæ. For Schmarda this and the two following groups rank only as families. *Hislopia* is an aberrant Indian freshwater form.

Catenicellidæ.
Catenicella.

Cellariidæ.
Salicornaria.
Cellaria.
Nellia.

Cellulariidæ.
Cellularia.
Menipea.
Scrupocellaria.
Canda.

Scrupariidæ.
Scruparia.
Hippothoa.
Ætea.
Beania.

Farciminariidæ.
Farciminaria.

Gemellariidæ.
Gemellaria.
Dimetopia.
Notamia.

Cabereidæ.
Caberea.

Bicellariidæ.
Bicellaria.
Bugula.
Naresia.

Flustridæ.
Flustra (Sea-mat).
Carbasea.

Membraniporidæ.
Membranipora.

Celleporidæ.
Cellepora.

Myriozoidæ.
Myriozoum.
Mollia.

Eschariporidæ.
Esch100.
Porina.
Anarthropora.

Escharidæ.
Eschara.
Lepralia.

Reteporidæ.
Retepora.

Vinculariidæ.
Vincularia.

Selenariidæ.
Selenaria.

Hislopia.

CYCLOSTOMATA.—Cells tubular, "partially free or wholly connate;" opening terminal, with a movable tip. No avicularia nor vibracula.

F 2

Two groups may be distinguished :—Articulata (= Radicata), zoarium divided into internodes ; Crisiidæ. Inarticulata (= Incrustata), zoarium not so divided ; including Frondiporidæ, with cells aggregated into bundles (Fasciculinea) ; the remaining families not so aggregated (Tubulinea). There are numerous fossil species.

Crisiidæ.	*Tubuliporidæ.*	Domopora.
Crisia.	Alecto.	Defrancia.
Crisidia.	Tubulipora.	
		Frondiporidæ.
Idmoneidæ.	*Diastoporidæ.*	Fasciculipora.
Idmonea.	Diastopora.	Frondipora.
Pustulopora.		
Hornera.	*Discoporellidæ.*	*Cerioporidæ.*
	Discoporella.	*Ceriopora.

CTENOSTOMATA.—Cell-opening closed by marginal setæ. No avicularia nor vibracula.

Only two families appear strictly to belong to this suborder, Alcyonidiidæ and Vesiculariidæ. *Pedicellina, Paludicella, Urnatella*, and *Loxosoma* are by some considered to represent as many suborders. Of these *Pedicellina* is sometimes referred to Gymnolæmata, sometimes to Phylactolæmata. Schmarda, adopting Nitsche's name of Endoprocta as an order, takes *Pedicellina* and the last two genera as the types of three families ; the rest of Polyzoa forming the order Ectoprocta.

Alcyonidium gelatinosum is *A. diaphanum* of Hooker's ' Flora Scotica.'

Vesiculariidæ.	Bowerbankia.
Vesicularia.	
Serialaria.	*Alcyonidiidæ.*
Valkeria.	Alcyonidium.

Order II. PHYLACTOLÆMATA.

LOPHOPODA.   HIPPOCREPIA.

Separation of individuals incomplete. Mouth with an epistome (or epiglottis). Lophopore bilateral, crescent-shaped, with numerous tentacles.

" The young, on its escape from the statoblast, is at first soli-

tary, but is rapidly multiplied by the production of gemmæ."
(*Allman.*)

These organisms are larger and more homogeneous than the
Gymnolæmata, aud have a soft or leathery or gelatinous structure, with no special stem. They are found attached to stones,
plants, &c. in fresh water, only one or two species being marine.
Cristatella mucedo ("not unlike certain hairy caterpillars," and
two inches in length) is found in many of our pools, creeping about
in a sluggish manner.

Rhabdopleura is an anomalous marine form from Norway; it
is said to have Hydrozoan affinities. Allman would place it
in a distinct "suborder," for which he suggests the name of
Aspidophora.

Plumatellidæ.	Lophopus.
Plumatella.	
Alcyonella.	*Cristatellidæ.*
Fredericella.	Cristatella.

Rhabdopleura.

Subkingdom V. ARTHROPODA.

CONDYLOPODA. ARTICULATA. GNATHOPODA.
ARTHROZOA.

Segmented, non-ciliated animals, with distinctly jointed legs. Nervous system ganglionic. Sexes generally separate. Oviparous.

The Arthropoda, with more than 200,000 species, vary to such an extent that little can be said applicable to the whole group. Of all Invertebrata they are the most advanced in the development of the organs peculiar to animal life, "manifested in the powers of locomotion, and in the instincts, which are so varied and wonderful in the insect class."

The Arthropoda fall naturally into four classes, which A. Agassiz, with cruel refinement, only ranks as orders.

Head, thorax, and abdomen distinct .... INSECTA.
Head or abdomen not distinct from the thorax.
    With antennæ.
      One pair .... MYRIOPODA.
      Two pairs .... CRUSTACEA.
    Without antennæ................................. ARACHNIDA.

Class I. CRUSTACEA.

AGONATA. BRANCHIOPNOA.

Mostly aquatic animals, with articulated appendages as well on the thorax as on the abdomen: Two pairs of antennæ in most.

The segments of Crustacea consist, at least theoretically, of a "tergum" of two pieces and an "epimeron" on each side above; and beneath of a "sternum," also of two pieces, and two lateral "episterna." The hard processes of the internal skeleton are the "apodemata," and serve for the attachment of muscles &c.

Some of the lower forms of Crustacea retrograde after passing the embryonic stage, but an advancing and gradual metamorphosis is more general. Three larval forms may be distinguished:—(1) "*Nauplius*," oval, unsegmented; one eye; three pairs of appendages, which are converted into antennæ and gnathites : (2) "*Zoëa*," elongate, segmented; thorax with a dorsal spine; two sessile eyes; abdomen as long as the body; legs rudimentary: and (3) "*Megalopa*," flattened, segmented, no dorsal spine; two pedunculate eyes; abdomen much diminished, partially bent under the body; five pairs of legs. There are intermediate forms, and, not to be referred to the above, are *Alima*, *Erichthus*, and *Phyllosoma*, which are also said to be larvæ—the two former of Squillidæ, the latter of Scyllaridæ. The lower Crustacea do not pass the *Zoëa*-stage; and some go through the *Nauplius*-stage in the egg.

The sexes sometimes differ considerably, and there are occasionally two forms of males. *Cymothoa* and some allied genera are said by Bullar not to have distinct sexes.

Owing to the extreme modifications in this class it is convenient to divide it into six subclasses. Some of these, however, and many of the minor groups, pass into one another with scarcely any line of differentiation. The grades of the divisions are therefore variously estimated by authors, and the names are not always conterminate.

Adult parasitic; if attached by the feet, .
  then the feet rudimentary.
    Sexes united............................. Cirripedia.
    Sexes distinct .... Epizoa.
Adult free, or, if parasitic, attached by
  the feet, the feet jointed.
  With respiratory organs.
    Eyes sessile.
      Carapace in one or two pieces... Entomostraca.
      No carapace....................... Edriophthalma.
    Eyes pedunculate .... Podophthalma.
  No respiratory organs................. Podosomata.

Subclass I. CIRRIPEDIA.

Pectostraca.

Body furnished with a mantle, and enclosed in a many-valved carapace. Abdomen rudimentary or obsolete. Feet in the form of cirri. Mostly hermaphrodite.

The adult is attached to some foreign body by the anterior extremity of the head, which is of large size, and almost always enveloped in a carapace containing also the rest of the body.

The posterior extremities consisting of a rudimentary abdomen, thorax, and six pairs of many-jointed limbs, which are used for capturing its food. They have no heart.

When the sexes are separate, the males are very minute and epizoic on the females; they are very rudimentary, mere sperm-sacs, and their characters are valueless for classification.

The metamorphosis of the Cirripedia is very complex, but in the earlier stages the larvæ resemble the Entomostraca (*Nauplius*, *Cypris*). Darwin divides them into three orders, to which Claus adds the Rhizocephala. Schmarda retains the latter in the Epizoa (Ichthyophthira).

Body unsegmented	RHIZOCEPHALA.
Body segmented.	
Without limbs	APODA.
With limbs.	
Limbs abdominal	ABDOMINALIA.
Limbs thoracic	THORACICA.

Order I. RHIZOCEPHALA.

SUCTORIA.

Body cylindrical or sac-shaped, without segmentation. No limbs, organs of sense, nor intestine.

There are two openings into the body, one serving as a mouth, the other as an anus. The reproductive organs are well developed.

These are sac-like or disciform parasites on crabs, to which they attach themselves by root-like tubes [modified antennæ] proceeding from the anterior portion of the body, penetrating and intertwining themselves into the substance of their victim. Crabs infested with *Sacculina* are barren, at first mechanically, afterwards histologically. *Peltogaster* is parasitic on the abdomen of Paguridæ. An opinion has been expressed that the *Peltogaster* of the *Pagurus* has become a *Sacculina* on the crab.

The young pass through a *Nauplius* and a *Cypris* stage.

Sacculinidæ.	*Peltogastridæ.*
Sacculina=Pachybdella.	Peltogaster.
Clistosaccus.	Apeltes.
Lernæodiscus.	Sylon.

Order II. APODA.

"Carapace reduced to two separate threads serving for attachment." Body without cirri. Mouth suctorial.

Proteolepas bicincta is the only member of this order. It is like the larva of an insect, about one fifth of an inch long, and parasitic on *Alepas cornuta*. Its earlier stages are unknown.

Proteolepas.

Order III. ABDOMINALIA.

" Carapace flask-shaped." Thoracic segments without limbs ; the abdomen with three pairs. Two eyes. Mouth extensile. Sexes distinct.

The members of this order are all burrowers in shells. *Cryptophialus minutus*, the only species of the genus known, is one tenth of an inch in length, and is lodged in a flask-shaped carapace. "The early larval stages are passed under an egg-like condition within the sac of the parent." The pupa, having no natatory limbs, crawls about by the aid of its antennæ. *Cochlorine* burrows in the shells of *Haliotis* ; *Alcippe* is found on our own coasts, in the shells of *Fusus* and *Buccinum*.

Cryptophialidæ.	*Alcippidæ.*
Cryptophialus.	Alcippe.
Cochlorine.	

Order IV. THORACICA.

"Carapace either a capitulum or a pedicel, or an operculated shell with a basis." Six thoracic segments with six pairs of limbs. Two eyes.

In this order are two primary forms, the pedunculate and the sessile. In the pedunculate forms the peduncle is formed by a modification of the larval antennæ. The sessile forms are protected by a strong multivalve conical shell closed by a four-valved operculum. " The whole shell has a cellular and organized texture, and its gradual expansion is provided for by the successive growth and calcification of processes of the mantle which penetrate the uniting sutures." The cement fixing the animal is secreted by an organ which Darwin has shown to be a modified portion of the ovarian tube.

The soft parts and cirri are subjected to a periodical moult, not, however, affecting the shell.

In the mature state the eyes only retain a certain susceptibility to light. The organs of hearing are two sac-like cavities situated at the base of the first pair of cirri.

The sexes are distinct in *Ibla Cumingii* and in *Scalpellum ornatum*; but many species of both genera are hermaphrodite, notwithstanding which they have also a (supplemental) male attached to them.

Besides many species found on, or burrowing into, whales, fish, mollusks, crabs, &c., one (*Ornitholepas australis*, a larval form, however) is said to attach itself to the feathers of a sea-bird (*Puffinus cinereus*).

There are about 100 species in this order.

Lepadidæ.

Anelasma.
Alepas.
Conchoderma =
    Otion.
Dichelaspis.
Pœcilasma.
Lepas (Barnacle).

Pollicipedidæ.

Scalpellum.
Ibla.

Lithotrya.
Pollicipes.

Coronulidæ.

Xenobalanus.
Tubicinella.
Coronula.

Balanidæ.

Chelonobia.
Pyrgoma.

Acasta.
Balanus (Acorn-
    shell).

Cthalamidæ.

Pachylasma.
Cthalamus.
Octomeris.

Verrucidæ.

Verruca = Clysia.

Subclass II. EPIZOA.

Suctoria. Ichthyophthira.

Body elongate, subarticulate, the antennæ and limbs terminated either by suckers, hooks, or bristles. Mouth suctorial. No respiratory organs. Females with external pendent ovisacs.

The Epizoa are deformed and grotesque ectoparasites of fish and other marine and freshwater animals. They differ from all other Crustacea, except Cirripedia, Copepoda, and Podosomata, in having no branchiæ. They attach themselves to their prey either by a suctorial mouth—a conical tube resulting in a modification of the upper and lower lips, accompanied by two bristle-shaped pieces, the analogues of the mandibles—or by a circular disk formed upon the confluent extremities of the posterior pair of feet. Another mode of adhesion is by certain processes that grow from the head.

The normal number of thoracic segments is five, but, in general, two or more are fused. The abdomen is terminated by two fin-shaped or setiform appendages, but is frequently rudimentary, or in some reduced to its two appendages. The ovisacs are attached to the last thoracic segment, where they remain, even after the ovaries have parted with their contents.

The males are mostly rudimentary, but of many species they are unknown. The young are free, and resemble Copepods ; but there is a metagenesis resulting in a usually permanent attachment to fishes, crustacea, or mollusks, to which they adhere in various ways. "Their development would seem to have been at first, as it were, hurried forward at too rapid a pace, and the young parasite, starting briskly into life, ranging to and fro by the highest developed natatory organs we have yet met with, and guiding its course by visual organs, must lose its eyes and limbs before it can fulfil the destined purpose of its creation." [*Owen.*]

The Epizoa are supposed to be more numerous than the whole class of fishes. They are the "Crustacés suceurs" of Milne-Edwards, but including also his 'Crustacés aranéiformes" [Podosomata]. By many modern zoologists they are included in the Copepoda, to which they are nearly allied ; but Schmarda (1878) continues to keep them apart.

Three or four pairs of limbs .. .. SIPHONOSTOMA.
Limbs rudimentary .... LERNÆODEA.

Order I. SIPHONOSTOMA.

PARASITA. THECATA. PŒCILOPODA. ONCHUNA.
CORMOSTOMATA.

Body divided into head, thorax, and abdomen ; the thorax segmented. Three pairs of short thoracic foot-jaws. "Two antennæ." (*Milne-Edwards.*)

The head is generally confounded with the first or first two thoracic segments. The antennæ are two- or more jointed ; the inner or lower pair are modified into hook-shaped and clasping organs, and are not recognized as antennæ by some authors. Besides the three pairs of foot-jaws, there are four pairs of natatory legs. In a few species the thorax is furnished with two appendages, resembling elytra.

In this and the following order, the families and genera are given after Claus. Corycæidæ are often referred to Copepoda.

Corycæidæ.

Corycæus.
Sapphirina.

Ergasilidæ.

Ergasilus.
Nicothoë.

Bomolochidæ.

Bomolochus.
Eucanthus.

Ascomyzontidæ.

Ascomyzon.
Asterocheres.

Caligidæ.

Caligus.
Trebius.
Elytrophora.
Euryphorus.
Dinemura.
Pandarus.

Læmargus.
Cecrops.

Dichelestiidæ.

Eudactylina.
Dichelestium.
Lamprogena.
Lernanthropus.
Cycnus.
Lonchidium = Kröy-
   eria.

The following family forms, according to Claus, the suborder Branchiura; but it is referred by Gerstaecker to Branchiopoda. *Argulus* lays its eggs, instead of carrying them about in ovisacs.

Argulidæ.

Argulus.
Gyropeltis.

Order II. LERNÆODEA.

Limbs simple tegumentary lobes, without articulations, and only serving to fix the parasite on its prey. Thorax not ringed.

The abdomen, with few exceptions, is rudimentary, and some are altogether without limbs. The antennæ when present are indistinctly jointed, and those of the inner pair are in the form of hooks or claws. In the young state they resemble Copepoda.

In this order the females attach themselves to the eyes, mouth, skin, and especially to the gills of fishes. The males are found lying under the abdomen of the female.

The Lernæodea were placed by Latreille with the intestinal worms; at the same time he recognized their similarity to the Siphonostoma. They are now generally included in one order or group, in which also Schmarda places Rhizocephala. It forms the two orders Cephaluna and Brachiuna of Owen.

Chondracanthidæ.

Chondracanthus =
   Lernentoma.

Lernæidæ.

Lernæocera.

Lernæa.
Lernæonema.
Pennella.

Lernæopodidæ.

Achtheres.

Basanistes.
Lernæopoda.
Brachiella.
Tracheliastes.
Anchorella.
Tanypleurus.

Subclass III. ENTOMOSTRACA.

Body furnished with a carapace, consisting of one or two shells, sometimes multivalve. Limbs jointed, setiferous. Branchiæ attached either to the limbs or to the oral appendages. Mouth mandibulate. Sexes distinct.

These are mostly minute and, with some exceptions, freshwater animals, very prolific, animal feeders, and very rarely parasitic. They undergo a moulting process as they grow, and some a metamorphosis. Parthenogenesis is not uncommon.

The Entomostraca of Milne-Edwards is confined to the two orders Ostracoda and Copepoda. Dana includes in it his three " suborders," Gnathostomata [Phyllopoda and Lophyropoda, the latter containing Ostracoda and Copepoda], Cormostomata [Pœcilopoda and Arachnopoda, the former comprising the Epizoa only], and Merostomata [Xiphura]. Von Hayek confines the Entomostraca to Epizoa and Copepoda ; Claus and Schmarda ignore the name. Excluding Epizoa, the term is here used in its original significance.

Branchiæ, when present, attached to the mouth.
    Head and thorax covered by a carapace ...... COPEPODA.
    Body enclosed in a two-valve shell .... OSTRACODA.
Branchiæ attached to the legs.
    Recent.
        Mandibles and maxillæ simple.
            Feet few, not foliaceous .... CLADOCERA.
            Feet many, foliaceous .... PHYLLOPODA.
        Mandibles and maxillæ terminating in foot-
            like appendages .... XIPHURA.
    Extinct.
        Body above three-lobed .... TRILOBITA.
        Body not lobed above .... EURYPTERIDA.

Order I. COPEPODA.

LOPHYROPODA. CROPHYROPODA. CYCLOPACEA. GNATHOSTOMATA. EUCOPEPODA.

Body covered by a single shell (carapace). Abdominal segments free. Four pairs of natatory legs. No branchiæ. Tail setiferous.

In addition to the four pairs of legs, there is also a rudimentary pair, but attached to the abdomen. The head is tolerably distinct, having one eye (occasionally two) and two pairs of antennæ, the latter sometimes differing considerably in the males, and one pair acting as a natatory organ. The mouth is mandibulate. The young are naupliiform. The female has one, generally two, pendent ovisacs.

The Copepoda are minute free-swimming Crustaceans, found both in the sea and in freshwater. One species, *Cetochilus septentrionalis*, forms the principal food of the southern whales. It also at times abounds on our coasts. Notodelphyidæ are commensals in the respiratory sac of Ascidians.

Claus places the Epizoa in the Copepoda; while Schmarda ranks them as one of the two families composing the order Lophyropoda (= Copepoda).

Cyclopidæ.
Misophria.
Oithona.
Cyclops.
Cyclopina.

Harpacticidæ.
Thalestris.
Westwoodia.
Canthocamptus.
Harpacticus.

Peltidiidæ.
Zaus.
Peltidium.
Alteutha.

Calanidæ.
Cetochilus.
Calanus.
Temora.
Diaptomus.

Pontellidæ.
Irenæus = Anomalocera.

Pontella.

Notodelphyidæ.
Notodelphys.
Notopterophorus.

Ascidicolidæ.
Ascidicola.

Buproridæ.
Buprorus.

Order II. OSTRACODA.

CROPHYROPODA. OSTRAPODA. CYPRIDACEA.

Body not ringed, enclosed between two shell-like valves, and terminated by a bifid tail. The inferior or second pair of antennæ natatory. Branchiæ attached to the oral appendages. Two or three pairs of thoracic feet.

The valves of the shell can be completely closed; when open it allows the play of the feet and antennæ. The abdomen is rudimentary. The eye is single (*Baird*), but this is due to the union of two; in Cypridinidæ there are two movable pedunculate eyes. There are no eyes in Halocypridæ (*Claus*). The young at once assumes the shape of the parent.

The Ostracoda are minute freshwater animals, very lively, swimming by the action of their antennæ. They do not carry their eggs [about 20 or 30] as in most Crustacea, but deposit them on foreign bodies, fixing them by a greenish filamentous substance.

Cyprididæ.	Ilyobates.	*Cypridinidæ.*
Bairdia.	Loxoconcha,	Cypridina.
Candona.	Paradoxostoma.	Asterops.
Notodromus.		
Cypris.	*Halocypridæ.*	*Polycopidæ.*
	Conchœcia.	Polycope.
Cytheridæ.	Halocypris.	
Cythere.		*Cytherellidæ.*
Polycheles.		Cytherella.

Order III. CLADOCERA.

DAPHNIACEA.

Head distinct; a bivalve carapace covering the rest of the body. Four to six pairs of feet carrying the branchiæ. Second pair of antennæ large, branched, acting as swimming-organs.

The carapace is composed of two valves, joined together on the back. The feet are foliaceous, and are scarcely organs of locomotion. The eye is single, and very large.

The Cladocera abound in fresh water, and are very prolific. They are more or less transparent, and have frequent moultings. "Ephippial eggs" are eggs with an additional covering, supposed to be the inner lining of the ovary. They appear to be produced in the winter.

For Gerstaecker the Cladocera are merely a tribe of a subsection of a section of his suborder "Branchiopoda genuina." (*Bronn's* Cl. Ord. Thier-Reichs.)

Podontidæ.	*Daphniidæ.*	Acroperus.
Podon.	Daphnia.	Pleuroxus.
Evadne.	Moina.	Chydorus.
	Bosmina.	
Polyphemidæ.		*Sididæ.*
Polyphemus.	*Lynceidæ.*	Sida.
	Lynceus.	Latona.
Leptodoridæ.	Eurycercus.	Daphnella.
Leptodora.	Alona.	

Order IV. PHYLLOPODA.

BRANCHIOPODA. ASPIDOPHORA. CERATOPHTHALMA.

Body divided into many segments, nearly all carrying a pair of foliaceous legs. Antennæ small, one or two pairs, not natatory. The legs vary from eleven to sixty pairs. The abdomen is many-jointed, terminating in two long setaceous appendages. The eyes are two, or sometimes three ; occasionally, as in *Branchipus*, they are pedunculate. Branchipodidæ have no carapace ; they swim on ·their backs. Apodidæ have a large shield-shaped carapace. In Estheriidæ the body is enclosed in a soft bivalve shell.

Apus cancriformis, two or three inches long, sometimes met with in this country, changes its skin twenty times in two or three months. The males were unknown a few years ago. The name ἄπους, given by Frisch in 1732, and adopted by later authors, is unwarranted. The family name is Apusidæ for Claus.

The sudden appearance of animals of this order is due to the latent vitality of their ova, which are only hatched under favourable conditions. They are mostly freshwater animals.

Apodidæ.	*Branchipodidæ.*	*Estheriidæ.*
Apus.	Artemia.	Limnadia.
Lepidurus.	Branchipus = Chirocephalus.	Estheria.

Order V. TRILOBITA.

PALÆADÆ.

Marine animals, often of large size, belonging to the Palæozoic period. Cephalic shield with two sutures dividing the median from the two lateral regions. Limbs rudimentary, lamelliform.

These extinct Crustaceans are believed to be allied to *Apus* and *Nebalia*. Their body is composed of from six to twenty segments, and the tail or postabdomen is bent under the thorax. The limbs are unknown. According to Schmarda, there are 1600 species ; he enumerates seven families. They have been found almost entirely in the Palæozoic rocks. Among the principal genera are :—

*Isoteles.	*Asaphus.	*Ogygia.
*Amphyx.	*Calymene.	*Paradoxides.
*Arges.	*Homalonotus.	*Trinucleus.

Order VI. XIPHURA.

Pœcilopoda. Xyphosura. Merostomata.

Head not distinct from the thorax, covered by a carapace. A styliform process or tail terminating the abdomen. Legs six pairs, surrounding the mouth.

The basal joints of the legs surrounding the mouth represent mandibles and maxillæ, the rest of the joints are ambulatory and prehensile. The young have no tail, which in the adult equals the rest of the body. There are two eyes, with two ocelli. The eyes are not faceted.

The embryo in its latest stage is said to resemble certain Trilobites. Only one recent genus is known, which existed also in the Oolitic period. The other fossil genera occur chiefly in the coal-formations.

> Limulus (King-crab).
> *Belinurus.
> *Cyclus=*Halicyne.

Order VII. EURYPTERIDA.

Head with two large marginal eyes and two median ocelli. Body of numerous free segments (12), all, except the first two, without appendages.

These were marine animals, often of large size, belonging to the Palæozoic period. They are allied to *Limulus* and Trilobita, and, according to Van Beneden, cannot be separated from the Scorpions. They had only one pair of antennæ. *Pterygotus anglicus* is known to the Scotch quarrymen as the "Seraphim."

Schmarda makes this order a pendant to Xiphura. The two are joined by Claus to Phyllopoda, which comprises, according to him, as suborders, Cladocera and Branchiopoda. Branchiopoda, however, is by other authorities made to include Phyllopoda as well as Ostracoda and Trilobita.

Woodward has enumerated sixty-two species.

*Eurypterus.	*Stylonurus.
*Pterygotus.	*Hemirhypis.

G

Subclass IV. EDRIOPHTHALMA

ARTHROSTRACA. TETRADECAPODA.

Eyes sessile. No carapace. Head, thorax, and abdomen distinct, the two latter segmented. Branchiæ more or less connected with the legs. Sexes distinct.

The two anterior pairs of legs [gnathopoda] are homologous to the two outer pairs of foot-jaws of the Decapoda. The young resemble the parent to a certain extent, and are for some time protected by the mother. The principal transformations take place in the egg.

The Edriophthalma and Podophthalma form the Malacostraca of the older authors. Dana includes Trilobita and Rotifera in his Edriophthalma. Schmarda adopts the three following orders of Latreille:—

Abdomen well developed.
    Respiration by thoracic vesicles.............. AMPHIPODA.
    Respiration by foliaceous limbs .... ISOPODA.
Abdomen rudimentary .... LÆMODIPODA.

Order I. LÆMODIPODA.

Abdomen rudimentary. Branchial vesicles attached to two or three thoracic rings. The first thoracic segments confluent with the head; the first two pairs of legs attached to this part.

The dorsal portion of the segments is entire. The female is furnished with abdominal laminæ for retaining the ova. *Cyamus* is parasitic, chiefly on whales. *Caprella* and *Proto* are sluggish inhabitants of our rocky tidal pools.

Læmodipoda are for Claus a tribe of Amphipoda.

Caprellidæ.	Podalirius.
Caprella.	
Proto.	*Cyamidæ.*
Protella.	Cyamus.
Cecrops.	

Order II. AMPHIPODA.

Branchiæ consisting of membranous vesicles attached to the

base of the legs. Thorax with six or seven free segments. Abdomen of seven segments. Tail natatory or saltatorial.

In this order the four anterior legs, including the gnathopoda, are directed forward, the three posterior backward; the five posterior legs [pereiopoda] are the true walking-legs. Behind the legs are three pairs of appendages [pleopoda], plumosely fringed; these are the swimming-organs. The terminal segment of the body is the telson, varying much in structure. The eyes are sometimes so slightly differentiated as to disappear after death. The largest species, from Lake Baikal, is five inches long.

Dana includes this and Isopoda in one order—Choristopoda. It comprises *Crevettina*! and *Hyperina* of Claus.

Dulichiidæ.
Dulichia.

Chelurida.
Chelura.

Corophiidæ.
Corophium.
Cyrtophium.
Podocerus.
Amphithoë.
Cerapus.
Siphonœcetes.

Orchestiidæ (Sand-fleas).
Talitrus = Orchestia.
Hyale = Nicæa
= Allorchestes.

Gammaridæ.
Atylus = Pherusa.
Dexamine.
Iphimedia.
Odius = Otus.
Œdicerus.
Leucothoë.
Probolium = Montagua.
Phoxus.
Urothoë.
Gammarus.
Mœra.
Melita.
Niphargus.
Lysianassa.
Anonyx.
Pontoporeia.

Vibiliidæ.
Vibilia.

Hyperiidæ.
Hyperia = Lestrigonus.
Tauria = Metœcus.
Cystisoma = Thaumops.
Themisto.

Phronimidæ.
Phronima.
Primno.

Typhidæ.
Oxycephalus.
Rhabdosoma.
Pronoë.
Typhis.

Order III. ISOPODA.

POLYGONATA.

No branchial vesicles; the respiratory organs placed beneath the abdomen, and modified in various ways. Body depressed. Tail well developed.

The head is almost always distinct from the thorax; the latter consists of seven segments bearing seven pairs of legs, all, the two anterior excepted, more or less conformable. The epimera of the dorsal portion of the segments are small or wanting. The young are somewhat different in form and with fewer limbs: in Arcturidæ they are carried for some time clinging on to the antennæ of the mother. In ordinary Isopoda they are retained for a time in a kind af pouch formed by the membranous plates at the base of the legs. The sexes are often more or less dissimilar.

In the terrestrial Oniscidæ the two or four anterior pairs of branchiæ are modified into lungs or air-vessels.

Some of the Oniscidæ are land animals, and are known as hog-lice, sows, &c. One of our English species, *Platyarthrus Hoffmanseggii*, is found in ants' nests. *Limnoria terebrans* is very destructive to submerged timber. *Æga spongiophila* resides in *Euplectella aspergillum*. *Bopyrus squillarum* is found commonly under the skin of prawns. *Liriope* is a parasite on *Peltogaster*, itself a parasite.

Tanaidæ.
Apseudes.
Tanais.
Rhoëa.

Anthuridæ.
Anthura.

Anceidæ.
Anceus = Praniza ♀.

Cymothoidæ.
Cymothoa.
Ceratothoa.
Eurydice = Slabberina.
Cirolana.
Æga.
Serolis.

Sphæromidæ.
Sphæroma.
Cymodocea.
Dynamene.
Campecopea.
Nesæa.
Monolistra.

Idoteidæ.
Idotea.
Chætilia.
Arcturus.
Leachia.

Munnopsidæ.
Munnopsis.

Asellidæ.
Asellus.

Limnoria (Gribble).
Iœra.
Munna.

Bopyridæ.
Bopyrus.
Ione.
Liriope.
Gyge.
Phryxus.

Oniscidæ.
Ligia.
Oniscus.
Porcellio.
Platyarthrus.
Philoscia.
Armadillo.
Tylos.

Subclass V. PODOPHTHALMA.

THORACOSTRACA.

Eyes on movable peduncles. Head and thorax confluent, covered by a shell or carapace.

The masticatory parts of the mouth [gnathites] are very complicated, the two or three anterior pairs of thoracic limbs being converted into foot-jaws [maxillipeds or pedipalps], and subject to considerable modifications. The tegumentary skeleton varies from being thin and flexible to a hard and solid calcareous crust.

In their development the Podophthalma vary greatly; in some reproduction is direct, but in others, and more generally, the young emerge from the egg as a *Nauplius* or a *Zoëa*, the *Nauplius*-stage, however, being sometimes passed in the ovum. A later larval form is the *Megalopa*. The changes are effected gradually, and while the animal is still comparatively minute; and they do not appear to be correlated with other characters.

Branchiæ external .... STOMATOPODA.
Branchiæ enclosed in the thorax .... DECAPODA.

Order I. STOMATOPODA.

STOMAPODA.   ANOMOBRANCHIATA.

Branchiæ external, either placed beneath the abdomen or attached to the thoracic legs, occasionally rudimentary or wanting.

The carapace covers the whole or only a part of the thorax, and is generally thin and flexible. The abdomen is elongate, and terminates in a natatory tail. The gnathites are confined to a pair of mandibles, two pairs of maxillæ, and a pair of foot-jaws, which are sometimes rudimentary, or are converted, as well as the seven succeeding pairs of limbs, into natatory feet.

The branchiæ consist of numerous minute cylinders, closely arranged on larger cylinders; they are wanting in *Mysis.*

Leuciferidæ compose the "tribe" Aplopoda of Dana. They are placed in the Macrura by Claus, who confines this order to the Squillidæ. The remainder are referred as a "suborder" to the Schizopoda of Latreille. *Nebalia*, formerly referred to Phyllopoda, is sometimes doubtfully placed here; it is a transition-form of a special type.

Mysidæ.	Euphausia.	*Squillidæ.*
Mysis.	*Lophogastridæ.*	Squilla.
Siriella.		Gonodactylus.
Petalophthalmus.	Gnathophausia.	Coronis.
	Lophogaster.	
Euphausiidæ.	Chalaraspis.	*Leuciferidæ.*
Thysanopoda.		Leucifer.
		Sergestes.

Nebalia.

Order II. DECAPODA.

Branchiæ enclosed in a special cavity on each side of the thorax. Five pairs of legs, the first pair didactyle.

The branchiæ consist of numerous thin plates placed closely together in the form of long quadrangular pyramids, nine on each side in the common crab, twenty-two in the lobster. The digestive organs comprise a large stomach and a large many-lobed liver. The heart is a contractile sac, with six main arteries. The eggs, after leaving the ovaries, are carried under the abdomen of the female, generally until they are hatched. In some, as in Gecarcinidæ, metamorphosis 'takes place within the egg.

The gnathites are composed of a pair of mandibles, two pairs of maxillæ, and three pairs of foot-jaws. There are five pairs of feet; the first and occasionally the second and third pairs are didactyle; the last pair is rudimentary in *Dynomene*. The segments of the head and thorax are closely soldered together and covered by the carapace.

Ecdysis, or moulting, occurs annually or oftener, until the animal ceases to grow. The muscles are previously subject to active absorption to within a third of their natural size in order to facilitate withdrawal. The animal escapes where the abdomen is connected with the carapace, or the carapace is split down the middle. Crabs begin to breed long before they attain their full size. They and their allies are the scavengers of the seas.

Milne-Edwards [1834] divided the Decapoda into three sections, two fairly natural, but the intermediate one very heterogeneous. In the first, Brachyura, the abdomen is small and folded under the body; in the third, Macrura, the abdomen is well-developed, in general longer than the body, with natatory appendages at the end. Anomura [forming, it is said, the passage between the two] has, if we except Paguridæ, little to distinguish it from one or the other; Claus suppresses the section, but places Porcellanidæ in the former. The Brachyura are further divided by Milne-Edwards into four families, the characters depending chiefly on the form of the carapace: thus, in the "Oxyrhinques" the carapace is narrowed anteriorly; in "Cyclométopes" broad and rounded anteriorly; in "Catamétopes" quadrilateral or ovoid; and in "Oxystomes" orbicular, prominent anteriorly. This is an artificial grouping; but the tribes into which he has divided them, and which are now ranked as families, are natural, and, with few exceptions, are adopted by recent authors. [French authors rank tribes as subordinate to the family.]

Our common crab is *Cancer pagurus* [closely allied species are found in North and South America and in New Zealand]; the shore-crab, *Carcinus mænas*; the spider crab or corwich, *Maia squinado*; the common land-crab, sold in the West-Indian markets, is *Cardisoma guanhumi*; but there are other species known as "land-crabs" belonging to *Uca, Gecarcinus*, &c.; the lobster is *Homarus gammarus*; the crawfish of the west of England and the London fishmongers is *Palinurus vulgaris* ["spiny lobster" seems to be a mere book-name]; the river crayfish, *Astacus fluviatilis*; the shrimp, *Crangon vulgaris*; and the prawn, *Palæmon serratus*. Two Mediterranean prawns are *Pandalus narwal* and *Penæus caramote*.

MACRURA.—Abdomen [or postabdomen] well-developed, the first five segments becoming natatory limbs, the sixth segment terminated in a five-parted tail-fin.

Claus heads Macrura with "Sergestidæ" (including *Leucifer*), Diastylidæ constituting the "suborder" Cumacea. Schmarda begins with Diastylidæ, *Leucifer* forming a pendant to Mysidæ and *Sergestes* ranking under Caridinidæ (=Alpheidæ).

Diastylidæ.
Leptocuma.
Leucon.
Cyrianassa.
Eudora.
Bodotria.
Diastylis=Cuma.

Penæidæ.
Stenopus.
Sicyonia.
Penæus.
Pasiphæa.

Palæmonidæ.
Hippolyte.
Palæmon (Prawn).
Pandalus.

Alpheidæ.
Alpheus.
Caridina.
Pontonia.
Athanas.

Crangonidæ.
Nica.
Lysmata.
Crangon (Shrimp).

Thalassinidæ.
Callianassa.
Axius.
Gebia.
Thalassina.

Astacidæ.
Astacus (River Crayfish).

Cambarus.
Homarus.
Nephrops.

Palinuridæ.
Palinurus (Crawfish).

Scyllaridæ.
Scyllarus.
Thenus.
Ibacus.

Galatheidæ.
Polycheles=Willemoësia.
Grimothea.
Galathea.
Munida.
*Eryon.

ANOMURA.—Abdomen slightly developed, except in Paguridæ, and without a tail-fin.

In this purely artificial division the two families Paguridæ and

Hippidæ are referred by Claus to Macrura, the remainder to the Brachyura; Porcellanidæ, Lithodidæ, Dromiidæ, and Dorippidæ forming his group "Notopoda;" while Raninidæ are placed, together with Leucosiidæ and Calappidæ, in the Oxystomata. Schmarda includes in *his* Anomura only the families represented by *Hippa*, *Lithodes*, and *Pagurus*. Here Milne-Edwards is provisionally followed, except that Dromiidæ change places with Dorippidæ.

Paguridæ.
Pagurus (Hermit-crab).
Cénobita.
Birgus (Tree-crab).

Porcellanidæ.
Porcellana.
Æglea.

Hippidæ.
Remipes.
Hippa.
Albunea.

Raninidæ.
Ranina.

Lithodidæ.
Lithodes.

Lomis.
Echidnocerus.

Homolidæ.
Homola.

Dorippidæ.
Cymopolia.
Dorippe.
Æthusa.

BRACHYURA.—Abdomen reduced to a triangular (male) or rounded (female) tail, lodged in repose in a depression in the breast, and without a tail-fin.

Schmarda adopts Milne-Edwards's four "families," with the addition of Dromiidæ. Claus has nineteen; excluding the Anomurous families, the principal difference in the following list is that he separates *Eriphia* from the Cancridæ, and unites Inachidæ to the Maiidæ. The last four families have been recently worked out by Miers, whose arrangement, reversing the order, is here followed.

Dromiidæ.
Dromia.
Dynomene.

Corystidæ.
Atelecyclus.
Thia.
Gomeza.
Corystes.
———
Telmessus.

Leucosiidæ.
Ebalia.
Philyra.
Oreophorus.
Nursia.
Ilia.
Ixa.
Leucosia.
Myra.
Persephone = Guaia.
Iphis.

Calappidæ.
Thealia.
Calappa.
Matuta.
Hepatus.
Orithyia.
Mursia.

Pinnotheridæ.
Pinnotheres.
Elamena.

Hymenosoma.
Myctiris.
Doto.

Grapsidæ.
Planes = Nautilo-
  grapsus.
Plagusia.
Helice.
Grapsus.
Brachynotus.
Sesarma.

Gonoplacidæ.
Gonoplax.
Macrophthalmus.

Ocypodidæ.
Ocypode.
Gelasimus.

Gecarcinidæ (Land-
  crabs).
Cardisoma.
Uca.
Gecarcinus.

Thelphusidæ.
Thelphusa.
Potamia = Boscia.

Portunidæ.
Podophthalmus.
Thalamita.

Lupa.
Platyonychus.
Portunus.
Polybius.
Carcinus (Shore-
  crab).
Portumnus.

Cancridæ.
Melia.
Pirimela.
Pilumnus.
Eriphia.
Etisus.
Cancer (Crab).
Zozymus.
Daira.
Ozius.
Chlorodius.
Xantho.
Atergatis.
Carpilius.

Parthenopidæ.
Zebrida.
Œthra.
Cryptopodia.
Parthenope.
Lambrus.

Periceridæ.
Mithrax.
Othonia.

Pericera.
Lissa.
Libinia.

Maiidæ
Micippa.
Cyclax.
Eurynome.
Hyastenus.
Pisa.
Maia (Spider-crab).
Herbstia.
Hyas.
Egeria.

Inachidæ.
Tyche.
Doclea.
Acanthonyx.
Epialtus.
Menæthius.
Huenia.
Chorinus.
Amathia.
Halimus.
Eurypodius.
Oncinopus.
Inachus.
Camposcia.
Achæus.
Stenorhynchus.
Leptopodia.

Subclass VI. PODOSOMATA.

Pycnogonida. Nymphonacea. Arachnopoda. Pantopoda.
Araneiformia. Aporobranchia. Polygonopoda. Levigrada.

No respiratory organs. Abdomen rudimentary, unsegmented.
Thorax of four segments, each carrying a pair of many-jointed
legs. Sexes distinct.

These are mostly spider-like, marine animals, very sluggish, and some of them external parasites. They have four ocelli, no antennæ, and a rostrate head, with a mouth either with or without mandibles and palpi. In *Nymphon* the digestive tube sends a branch into each limb, which are, according to Milne-Edwards, the seat of a peristaltic motion. Johnstone asserts that the heart also sends a branch into each limb. The females are known by having a pair of spurious legs for the purpose of carrying the eggs.

The earlier states are not well understood; but there appears to be first a naupliiform state, the larva afterwards losing its three anterior pairs of appendages. [In the Crustacean *Nauplius* these three pairs always represent the antennary and mandibular appendages.]

From their tegumentary respiration, having no tracheæ or pulmonary sacs, Milne-Edwards considers that they have more analogy with the lower forms of Crustacea, which are similarly conditioned. A. Dohrn denies their being either Arachnida or Crustacea. They are now generally placed with the Arachnida. Häckel at one time combined them with Arctisca to form his subclass "Pseudarachna;" but the latter he now places with the Worms; the Podosomata he retains in the Crustacea.

Pycnogonidæ.	*Nymphonidæ.*	Ammothea.
Pasithoë.	Nymphon.	Phoxichilidium=
Phoxichilus.	Zetes.	Orithyia.
Pycnogonum.	Achelia.	Pallene.

Oomerus.

Class II. MYRIOPODA.

MITOSATA.

Head distinct; thorax and abdomen not differentiated, divided into numerous segments [somites]. Two antennæ. Feet always more than eight pairs in the adult. Respiration tracheal. No metamorphosis.

Although there is no metamorphosis, the young have fewer segments, and three to six pairs of legs or none, but with each successive moult the number is increased. The mouth is complex, two pairs of feet sometimes enter into its composition; the mandibles are jointed. The eyes are simple or compound, in one or two rows, but sometimes absent. The females are largest, and are oviparous or ovoviparous. The respiratory, digestive, and nervous systems closely resemble those of the larvæ of insects.

Fossil species are found as early as the Carboniferous epoch.

The species of this class are known as gallyworms and centipedes. For Latreille they formed the first order of insects.

Body segmented.
    One pair of legs to each somite......... CHILOPODA.
    Two pairs of legs to each somite ...... CHILOGNATHA.
Body unsegmented .... MALACOPODA.

Order I. CHILOGNATHA.

DIPLOPODA.

Body more or less cylindrical and crustaceous. No foot-jaws. Antennæ of rarely more than seven joints. Two pairs of legs to each somite, except the first five or six.

The mandibles are without palpi; the second pair of gnathites (maxillæ) are united to form a lower lip. The number of somites varies from nine to eighty.

These are sluggish animals, living on decomposing animal and vegetable matter, and laying in the earth a great number of eggs. Glomeridæ are capable of rolling themselves into a ball. Polyzoniidæ are the Siphonizantia or Sugentia of Brandt.

A minute centipede, *Pauropus Huxleyi*, in its earliest stage with three pairs of legs, gradually increasing with each moult to nine pairs, was discovered by Sir J. Lubbock in Kent. He considers it the "connecting-link" between Chilognatha and Chilopoda; Packard places it between Myriopoda and Collembola. Claus puts it with Polyxenidæ. It differs from all other Myriopoda in having no tracheæ, and also in having bifid antennæ. Another species is found in North America.

Glomeridæ.	*Polydesmidæ.*	*Iulidæ.*
Glomeris.	Polydesmus.	Spirostreptus.
Zephronia.	Craspedosoma.	Iulus.
Sphærotherium.		Spirobolus.
	Polyxenidæ.	Lysiopetalum.
Polyzoniidæ.	Polyxenus.	
Polyzonium.		
Siphonophora!		*Archiulidæ.*
Brachycybe.	*Pauropodidæ.*	*Archiulus.
	Pauropus.	

Order II. CHILOPODA.

SYNGNATHA.

Body flattish, submembranous. Two anterior pairs of legs modified into foot-jaws. Antennæ with fourteen or more joints. One pair of legs to each somite.

The two mandibles have each a palpiform appendage; the second pair of foot-jaws are perforated for the discharge of a poisonous secretion. The eyes are generally numerous in the adult, and in *Cermatia* they are large and faceted; they are wanting in Geophilidæ. Some species of *Scolopendra* are said to be viviparous.

Except the Geophilidæ, these centipedes are very active and voracious; the bite of the larger species is highly venomous and very painful, leaving a callus which may last for months. Some of the *Geophili* have the property of secreting a phosphorescent matter.

Geophilidæ.	*Lithobiidæ.*	Eucorybus.
Strigamia.	Lithobius.	Scolopendra.
Cryptops.		
Geophilus.	*Scolopendridæ.*	*Cermatiidæ.*
Himantarium.	Heterostoma=Der-	Cermatia=Scutigera.
Arthronomalus.	cetum.	

Order III. MALACOPODA.

ONYCHOPHORA. PERIPATIDEA.

Body soft, cylindrical, unsegmented. Jaws foot-like, terminated by two curved claws. Legs from fourteen to thirty pairs. Viviparous. Sexes distinct.

There are two simple eyes and two tentacular-like antennæ. The lips are soft, and the mouth has a perforated papilla on each side. The legs are indistinctly articulated, and provided with two terminal claws. The tracheal pores are diffused over the surface of the body.

The body is unsegmented according to Huxley, but it has from 13 to 36 segments according to Schmarda. There are evident traces of segmentation in *Peripatus Edwardsii*, but there are none in *P. Blainvillei, P. juliformis, P. capensis*, and *P. novæ-zelandiæ*.

The members of this order, confined to a single genus, are found in the West Indies, South America, the Cape, Australia, and New Zealand. Hutton says of the New-Zealand species that it is nocturnal, living in decayed wood and under stones, feeding " upon animals," and able to shoot from its oral papillæ a viscid fluid, which hardens into a spider-like web, and by means of which it catches its prey. It breeds all the year round, although in the winter it is half-torpid.

These animals have been classed with the Vermes; but Moseley has shown that they really belong to the Arthropoda, in which they have also been placed by Schmarda, who has adopted Blainville's earlier name of Malacopoda.

Peripatidæ.
Peripatus.

Class III. ARACHNIDA.

Unogata. Acera.

Head united to the thorax (cephalothorax); no antennæ. Abdomen not segmented, or if segmented not distinctly separated from the cephalothorax, and never provided with limbs; legs never more than four pairs.

The palpi, mandibles, and sometimes the anterior pair of legs are variously modified; the latter are, according to Claparède, homologous with the labial palpi of insects; the maxillary palpi are known as "pedipalpi." All the appendages of the mouth being posterior, there are no homologues of the antennæ. The eyes are simple and generally more than two. Respiration is effected either by pulmonary sacs or by tracheæ, or by both, and more rarely by the skin alone. All Arachnida are digitigrade. Like the Crustacea they have the power of reproducing lost limbs.

Arachnida " occur in the Mesozoic formations, while spiders and scorpions of large size have been found in the Carboniferous rocks." According to Schmarda there are about 4600 species.

Solpugidea, Phalangidea, and Cheliferidea are sometimes united to form the order Adelarthrosomata; Phrynidea and Scorpiodea form another order—Pulmonaria, or they are all united under the name of Arthrogastra, Araneidea and Acaridea forming the Sphærogastra. Owen has also an order, Dermophysa, including Arctisca, *Demodex*, and Podosomata, characterized by the absence of distinct respiratory organs. Arctisca and Pentastomidea are two very aberrant and dissimilar groups.

Abdomen segmented.
    Respiration by tracheæ.
        Abdomen distinctly separated from the
            cephalothorax .... SOLPUGIDEA.   -
        Abdomen not distinctly separated.
            Maxillary palpi didactyle .... CHELIFERIDEA.
            Maxillary palpi monodactyle.............. PHALANGIDEA.
    Respiration by pulmonary sacs.
        A postabdomen terminating in a claw ...... SCORPIODEA.   -
        No postabdomen .... PHRYNIDEA.
Abdomen unsegmented.
    Abdomen united to the cephalothorax......... ACARIDEA.   -
    Abdomen distinct from the cephalothorax ... ARANEIDEA   -

A vermiform body with four pairs of rudimen-
    tary limbs .... ARCTISCA.
A vermiform body, the embryos only with two
    pairs of rudimentary limbs .... PENTASTOMIDEA.

Order I. SCORPIODEA.   (Scorpions.)

Abdomen indistinctly separated from the cephalothorax, of seven segments, with a six-jointed postabdomen having a curved claw on the extremity of the last segment. Maxillary palpi (pedipalpi) longer than the feet, terminating in a didactyle hand with a movable finger.

The head is furnished with a pair of cheliceræ or mandibles (supposed by some to be modified antennæ), each having a movable and a fixed claw. There are from six to twelve ocelli, two of which are larger and approximate, and placed more or less in the centre of the cephalothorax. The respiratory organs "are four pairs of flattened sacs, which open externally by as many stigmata, on the sterna of the four posterior free thoracic somites."

The curved claw or telson at the extremity of the abdomen is pierced for the passage of a venomous fluid, from a gland placed in the last segment. In both sexes there are peculiar comb-like organs situated directly behind the last pair of legs; their use is unknown. Scorpions are viviparous, producing 20–60 at a time; the young are carried on the back of the female for about a month. They live on insects, which they sting to death.

The Scorpions closely resemble in many respects the Xiphura and Eurypterida, from which Van Beneden thinks they cannot be separated. Fabricius placed them with the Crustacea in his order Agonata (Syst. Ent. 1775). They date back to the Carboniferous period.

This is such a homogeneous order that some authors recognize only a single genus. Peters, however, divides it into four groups, depending on the form of the sternum and the armature of the mandibles; more recently, and on different principles, Thorell makes four families and 31 genera. · With Phrynidea they form the Polymerosomata or Pedipalpi of some writers. Combining *Vejovis* with Scorpionidæ (Peters), and *Centrurus* with Androctonidæ (Thorell), we have three families.

Scorpionidæ.	*Telegonidæ.*	*Androctonidæ.*
Iurus.	Bothriurus.	Centrurus.
Heterometrus.	Telegonus.	Lychas = Tityus.
Scorpio.	Cercophonius.	Isometrus.
Vejovis.		Buthus.
Ischnurus.	——	Androctonus =
	*Cyclophthalmus.	Prionurus.

Order II. CHELIFERIDEA.

PSEUDOSCORPIONES.

Abdomen segmented, indistinctly separated from the cephalothorax, mutic at the extremity. Maxillary palpi longer than the feet, terminating in a didactyle hand with a movable finger. Eyes 2–4.

Like the Spiders the Cheliferidea are provided with silk-glands, and unlike the Scorpions, which they externally resemble, they have neither a postabdomen nor poison-glands. They breathe by tracheæ.

These Arachnids are of small size, and are found chiefly in caverns and damp places in temperate countries. *Chelifer cancroides* is often to be met with among old books. Owen's order Trachearia comprises this and three co-ordinate groups, "Acarina, Opilionina, and Solpugii."

Obisiidæ.	*Cheliferidæ.*
Cthonius.	Chernes.
Blothrus.	Garypus.
Obisium.	Chelifer.
Roncus.	Olpium.

Order III. ACARIDEA.

MONOMEROSOMATA.

Head, thorax, and abdomen united. Eight legs, six in the young. Respiration tracheal or dermal. With or without eyes. Mouth either masticatory or suctorial.

The bases of the cheliceræ and of the pedipalpi sometimes coalesce with the labrum and give rise to a suctorial rostrum. In some mites certain pairs of legs are terminated by suckers, in others by setæ. In a common form, probably the earlier state of a *Trombidium* (*Acarus libellulæ* of Carus), of a bright scarlet colour, found attached to the wings of various insects, there are no legs or other appendages. *Phytoptus* (supposed by some to be a larval form) does not appear to have more than four legs.

Acaridea are mostly oviparous; some are subject to a kind of metamorphosis, being worm-like when hatched. They are generally parasitic, but many are also found in dung, decaying matter, and on plants. Some are marine, and a few are found in fresh water. *Acarus domesticus* is the cheese-mite. *Leptus autumnalis* is the harvest-bug. *Gamasus coleopterorum* occurs abundantly on dung-beetles. *Atax* lives in the branchiæ of *Mytilus*, *Unio*, and *Anodonta*; and *Halarachne halichœri* in the nostrils of the seal. *Demodex folliculorum* is found in the sebaceous follicles of man; in the dog it causes the mange. It is a very minute animal, footless, and without a mouth; after two or three changes of skin it acquires its adult condition. The itch is caused by *Sarcoptes scabiei*, an eyeless form, burrowing through the skin.

Various species of *Phytoptus* are very injurious to vegetation, either through the bud or through the leaves; they browse on the surface of these " until they have flayed it to the quick." Galls are frequently caused by them; but these are not true galls, as " they have always an opening leading into them " (*Murray*). The " witch-knot " found on the birch, and resembling a great mass of twigs like a bird's-nest, is an abnormal growth of some years caused by *Phytopti*. Some of these galls were formerly supposed to be Fungi. and received from botanists such names as *Erineum*, *Phyllarium*, &c.

Tetranychus telarius (the red spider) spins silky webs on the leaves of various plants. Its colour is very variable.

The young Acaridea with six feet were placed by Latreille in his family " Microphthires." Hermann combined this order with Phalangidea under the name of " Holetra."

Demodicidæ.
Demodex = Simonea.

Sarcoptidæ.
Sarcoptes.
Dermalichus = Anal-
   ges = Myocoptes.

Acaridæ.
Acarus = Tyrogly-
   phus.
Glyciphagus.
Myobia.

Gamasidæ.
Gamasus.

Dermanyssus.
Pteroptus.

Ixodidæ.
Argas.
Ixodes.
Amblyomma.

Trombidiidæ.
Tetranychus.
Leptus.
Linopodes.
Erythræus.
Trombidium.
Phytoptus.

———

Pœcilophysis.

Hydrachnidæ.
Limnochares.
Hydrachna.
Hydrochoreutes.
Atax.

Oribatidæ.
Halacarus.
Oribates = Notaspis.
Nothrus.
Hoplophora.
Damæus.

Bdellidæ.
Bdella.
Scirus.

Order IV. ARANEIDEA. (Spiders.)

DIMEROSOMATA. PULMOTRACHEARIA.

Respiration by pulmonary sacs, often combined with tracheæ. Abdomen distinctly separated from the cephalothorax, not segmented. Two palpi. Eyes simple, 4-8. Eight feet, each of seven joints.

There are special abdominal glands for the secretion of a gummy fluid, which, exposed to the air, hardens into silk, used for the construction of webs; they communicate externally with the spinnerets by long convoluted vessels. There are also special glands for the secretion of a poison, placed at the base of the falces or poison-fangs; these are situated between the eyes and the mouth, and are hollow for the passage of the fluid. The mouth has two maxillæ, each of which bears a palpus, often leg-shaped. In the male spiders the parts analogous to the vesiculæ seminales are lodged in these maxillary palpi, but the testes are, as usual, in the abdomen. The males are much smaller and weaker than the female, and are frequently sacrificed to her savage instincts—*etiam in amoribus sæva.* They moult when growing, but there is no metamorphosis. In the embryonic state there is a postabdomen which gradually disappears. *Arthrolycosa,* a fossil

H

species from the Coal-measures of Illinois, appears to have had a segmented abdomen.

Spiders are oviparous, generally placing their eggs in a silky cocoon, which is sometimes carried by the mother beneath the body. Species of *Cteniza, Atypus*, and other genera construct tubular burrows, lined with silk or otherwise, and closed with a hinged door. These are the "trap-door spiders." Gossamer is the web formed by young spiders, by which they are enabled to float to a great height in the air.

Argyroneta aquatica is the well-known water-spider; the house-spider is *Tegenaria domestica*. The "bird-spider" (*Mygale avicularia*) is strong enough to kill small birds.

Spiders have been divided into two groups dependent on the number of pulmonary sacs [Dipneumones with two, Tetrapneumones with four]; also on the number of eyes [two, six, or eight]. Only one is known with two eyes, *Nops guanabacoæ*, a Cuban spider; three or four have four eyes. The species with two pulmonary sacs, the great bulk of the order, have been formed into two divisions, Sedentariæ and Vagabundæ, the former again divided into Orbitelariæ or Inæquitalæ, Retetelariæ, Tubitelariæ, and Laterigradæ; the latter into Citigradæ and Saltigradæ; while the species with four pulmonary sacs comprise only one division, Territelariæ. These divisions, however, dependent chiefly upon habits, form of the web, &c., seem incapable of being satisfactorily defined; and accordingly they have only been indicated in the following list, in which Thorell's arrangement has been generally adopted. Recently Bertkau proposes to divide spiders according to the number of spiracles :—Tetrasticta with two pairs, Tristicta with one pair; the latter including all the families except "Atypidæ" and Dysderidæ.

According to Cambridge, there are about 500 British species of spiders.

ORBITELARIÆ.	*Gastracanthidæ.*	RETETELARIÆ.
Epeiridæ.	Gastracantha.	*Theridiidæ.*
Tetragnatha.	Acrosoma.	Theridion.
Epeira.	Cyrtarachne.	Neriene.
Argyope.		*Mizalia.
Nephila.	*Arcyidæ.*	Erigone.
Miagrammopes.	Arcys.	Walckenaëra.
Uloborus=Veleda.	*Archæa.	Linyphia.
		Argyrodes.
		Tapinopa.
Phoroncidia.	Poltys.	Steatoda.

Scytodidæ.
Pholcus.
Latrodectus.
Scytodes.

Enyidæ.
Enyo.

TUBITELARIÆ.
Agelenidæ.
Argyroneta.
Amaurobius = Ciniflò.
Dictyna.
Lethia.
Agelena.
Tegenaria = Aranea.
Uroctea = Clotho.
Œcobius.
Hersilia.
*Gerdia.
Cœlotes.

———

Tetrablemma.

Drassidæ.
Hecaerge.
Liocranum.
Drassus.
Clubiona.
Gnaphosa.

Dysderidæ.
Stalita.
Nops.
Segestria.
Harpactes.
Dysdera.

Filistatidæ.
Filistata.

TERRITELARIÆ.
Mygalidæ.
Mygale = Theraphosa.
Atypus.
Cteniza.
Eriodon.
Idiops.
Xysticus.
Nemesia.
Anetes.
*Phalangites.

Liphistiidæ.
Liphistius.

Catadysidæ.
Catadysas.

LATERIGRADÆ.
Thomisidæ.
Stephanopis.
Sparassus.
Misumena.
Diæa.
Thomisus.
Oxyptila.
Thanatus.
Philodromus.
Selenops = Hypoplatea.
*Chalinura.

———

Heteropoda =
Sarotes.

CITIGRADÆ.
Lycosidæ.
Lycosa.
Tarentula.
Pirata.
Dolomedes.
Ctenus.
Ocyale.

Oxyopidæ.
Sphasus.
Oxyopes.
Peucetia = Pasithea.

———

Podophthalma.

SALTIGRADÆ.
Eresidæ.
Eresus.
Chersis = Palpimanus.

———

Dinopis.

Salticidæ.
Salticus.
Euophrys.
Marpessa.
Attus.
Ælurops.
Synemosyne.

———

Otiothops.

Myrmeciidæ.
Myrmecia.

———

*Arthrolycosa.

*Protolycosa.

Order V. PHALANGIDEA.

OPILIONINA.

Abdomen segmented, indistinctly separated from the cephalo-thorax. Maxillary palpi filiform, with a single claw. Eyes two. Respiration by tracheæ.

Unlike spiders, they have no poison-gland or spinning-organs; and it is difficult to tell the males from the females, except from the greater length of the maxillary palpi; but the latter have an ovipositor. The young have the same form as the adult. *Dolichoscelis Haworthii* has legs 25 times longer than the body. The Phalangidea are very voracious and destroy one another. The British species are sometimes called "harvest-spiders;" they all, with one rare exception, belong to the family Phalangiidæ. Unlike spiders, they die, so far as is known, at the end of the autumn.

Siro has the eyes widely apart, each placed on a peduncle.

Cryptostemmidæ.
Cryptostemma.

Sironidæ.
Siro=Cyphophthal-mus.

Gonyleptidæ.
Goniosoma.

Scotolemon.
Gonyleptes.
Stygnus.
Eusarcus.

Trogulidæ.
Trogulus.

Phalangiidæ.
Egænus.

Ischyropsalis.
Discosoma.
Leiobunum.
Opilio.
Phalangium.
Dolichoscelis.

Cosmetidæ.
Cosmetus.

، Order VI. PHRYNIDEA.

Abdomen segmented, distinct from the cephalothorax, with or without a seta or style at the end. Palpi long, leg-shaped, mono-or didactyle. Anterior pair of legs simulating antennæ.

There are 4–8 pulmonary sacs. The ocelli (or eyes) are ordi-narily eight; none, however, in *Nyctalops*.

In Phrynidæ the abdomen is rounded and mutic, and the maxillary palpi are armed with a single claw. In Thelyphonidæ the maxillary palpi are didactyle, and the abdomen is furnished with a jointed setiform appendage.

These Arachnids are all tropical, living under stones and in damp places; they are not venomous. Latreille combined them with the preceding to form his family Pedipalpi. *Tarantula* of Fabricius included *Phrynus* and *Thelyphonus.*

Phrynidæ.	*Thelyphonidæ.*
Phrynus.	Thelyphonus.
*Architarbus.	Nyctalops.

Order VII. SOLPUGIDEA.

Abdomen segmented, distinct from the cephalothorax, mutic at the extremity. Palpi filiform, simple, porrect. Two eyes.

The mandibles, modified antennæ, are very large, didactyle, and without a poison-gland. The palpi are in the form of legs or of antennæ, and, like the first pair of feet, are without claws. The body is hairy.

These Arachnids are found in the hot countries of the Old World. They are very ferocious; *Galeodes* will attack and kill small mammals, biting into them until its jaws have attained a vital part.

Galeodidæ.

Galeodes = Solpuga.	Rhax.	Gluvia.
Aellopus.	Gætulia.	Datames.

Order VIII. ARCTISCA.

MACROBIOTIDA. TARDIGRADA. COLPODA. CORMOPODA.

Minute, worm-like animals, with eight very short, indistinctly triarticulate feet. Cephalothorax and abdomen united.

The mouth is suctorial, with two styles, the rudiments of lateral jaws. There are no respiratory organs nor heart. The last pair of legs is given off from the abdomen. They are mostly hermaphrodite.

The Arctisca are known as "water-bears;" they are very slow in their movements, and are found in damp places. They were originally grouped with the Rotifers, and, like them, they regain their vitality after being desiccated, and apparently dead, for many years. They are oviparous; the eggs are very large, and the young have the same number of legs as the adult.

Macrobiotidæ.

Macrobiotus.
Emydium = Echiniscus.
Milnesium.

Order IX. PENTASTOMIDEA.

LINGUATULINA. ACANTHOTHECA.

Worm-like entozoic animals without feet, but the embryo with four rudimentary legs. Body long, annulated. Sexes distinct.

These are colourless parasites, classed by Rudolphi with Trematoda, and having a parasitism very similar to the Cestoda. They have no special organs for respiration. The mouth has two pairs of hooks in lieu of limbs. They require about a year to attain the adult condition.

These parasites are found in a sexless condition in the lungs and liver of hares and other herbivorous mammals and of reptiles; and in the sexual state in the nostrils of dogs and other carnivora by whom the herbivorous mammals have been devoured. The males are much smaller than the females; the latter in *Pentastoma tænioides* is 3 or 4 inches long. The larvæ of *P. constrictum* has been found encysted in the human liver. There is only one genus, with upwards of twenty species.

Pentastoma.

Class IV. INSECTA.

CONDYLOPODA. HEXAPODA.

Head, thorax, and abdomen distinct. Two antennæ. Three pairs of legs. Somites never more than twenty. Respiration by tracheæ. Sexes distinct.

The wings, which are almost always present, are developed from the second and third thoracic somites. They "are essentially flattened vesicles, sustained by slender but firm hollow tubes called nervures [but more analogous to veins], along which branches of the tracheæ and channels of circulation are continued." By Oken they were called "aerial gills," the homologues of the tergal branchiæ of the Vermes.

The eyes are either simple or compound; the former (ocelli or stemmata), situated on the vertex, are generally three in number, and are found in all orders; compound eyes, always two, but in rare cases divided or apparently so, are made up of a number of cones, separated from one another by a layer of pigment, the external broad end hexagonal, capped by a facet or "corneule," the narrow end communicating with the optic nerve. These facets vary in number; there are about 40 in *Myrmecina La-*

treillei; in *Mordella*! (120 species) there is said to be 25,000; in the largest beetle, *Dynastes Hercules*, they are so numerous as only to be seen by the most powerful lens. Only the rays of light which enter the cones in a direct line can reach the optic nerve, and yet insects have the power of discerning objects at comparatively great distances. The larvæ of insects having a complete metamorphosis are said to be destitute of compound eyes. Besides the antennæ there are a pair of mandibles, often aborted and two pairs of maxillæ, the second pair coalescing and forming the labium; but either pair may become suctorial, and of the labium only the palpi may be developed. In all orders there are species with the mouths either obsolete or rudimentary.

In all insects there are never more nor less than six legs, but the fore legs are sometimes rudimentary. The leg consists of five parts—coxa, trochanter, femur, tibia, and tarsus. The coxa is articulated to the thorax, the trochanter is attached both to the coxa and the femur: the tarsus is made up of from one to five joints, and almost invariably of a pair of claws; it is either naked, hairy, or scaly beneath; the joints are sometimes furnished with a dilatable membranous sac.

Of somites the normal number is thirteen, one for the head, three for the thorax, and nine (or, according to some, eleven) for the abdomen; the head, however, is assumed to consist of from four to six "coalesced somites," each somite being indicated by a pair of appendages.

The nervous system consists of ganglia (varying from one to eleven) connected by double commissures, giving off the nerves. The supracœsophageal ganglion is sometimes called the brain, with the functions of which it is analogous; the subcœsophageal ganglion supplies the mouth and its appendages.

Four or more slender cylindrical vessels, terminating at the commencement of the small intestine, are supposed to perform the function of a liver. They are known as "Malpighian tubes," and are found in most insects. The heart is a long dorsal tube, largest in the abdomen, where it is included in a saccular venous sinus from which the colourless blood passes into it, and, flowing towards the head, returns to the sinus by certain lacunæ. There are no arteries or veins. The tracheæ are aeriform tubes; they commence from lateral openings, known as "stigmata" or "spiracles," and ramify over every part of the body.

Nearly all insects undergo a metamorphosis, which may be complete or incomplete. The pupa, when quiescent, is either entirely enclosed in its case, or each limb may be more or less separately enclosed; when the pupa is active it may resemble the imago (only wanting the wings, or the wings only partially developed), or it may have no resemblance to it.

Agamic reproduction (parthenogenesis) occurs in some of the orders, chiefly among the Hemiptera and Hymenoptera; this is said to be the result of cell-formation, " comparable to a kind of budding."

The life of an insect for good or for evil, with a few exceptions, is in its larval stage; in its perfect state its life is short, sometimes lasting only a few hours. In the fertilization of plants insects play an important part by conveying the pollen from the anthers to the stigma.

Insects and their allies are supposed by Herbert Spencer to be compound animals, each one representing as many individuals as there are true segments, but which have become severally specialized for certain definite functions.

Considering the Mallophaga to be degraded forms of Hemiptera, and that the Trichoptera are not more entitled to ordinal rank than other groups of Neuroptera, there will remain nine orders which, if we except Collembola and Thysanura, are universally admitted. To these are sometimes added Thysanoptera [aberrant Hemiptera], Euplexoptera [an isolated form of Orthoptera], Aphaniptera [degraded Diptera], and Strepsiptera and Achrioptera [abnormal Coleoptera].

Apterous: no metamorphosis (Ametabola).
  Abdomen with terminal saltatory appendages... COLLEMBOLA.
  Abdomen without such appendages............... THYSANURA.
Winged: metamorphosis in nearly all.
  Metamorphosis incomplete (Hemimetabola, Homomorpha).
    Without mandibles [suctorial].................. HEMIPTERA.
    With mandibles.
      Wings dissimilar in texture .... ORTHOPTERA.
      Wings similar in texture .... NEUROPTERA.
  Metamorphosis complete (Metabola, Heteromorpha).
    Without mandibles.
      Two wings .... DIPTERA.
      Four wings .... LEPIDOPTERA.
    With mandibles.
      Upper wing coriaceous .... COLEOPTERA.
      Upper and lower wings membranous ...... HYMENOPTERA.

After the Collembola and Thysanura, the Hemiptera contain

some of the lowest forms of insect-life, and Hymenoptera probably the highest; between these it is scarcely possible to arrange the remaining orders in any thing like a satisfactory manner. There cannot be said to be any connecting links between them, although superficial resemblances may in a few instances be observed.

Häckel, according to his views of the succession of insect-life on the earth, places the orders in this sequence:—Archiptera, Neuroptera, Orthoptera, Coleoptera (these orders have mandibulate mouths, and were the only insects existing for a long period; the last three, he continues, were probably derived from the Archiptera, i. e. *Ephemera, Libellula, Lepisma, Termes,* &c.), Hymenoptera, Hemiptera, Diptera, and Lepidoptera. An ancient fly (*Eugereon*) in the Permian system seems to indicate, he thinks, the derivation of the Hemiptera from the Neuroptera. Gerstaecker (1873) begins with the Orthoptera and ends with the Hemiptera, placing Hymenoptera in the middle as the highest type of the insect-forms.

Fritz Müller thinks " that the most ancient insects " approached most nearly to the wingless Blattidæ, and that the complete metamorphoses of Diptera, Lepidoptera, Coleoptera, &c. were of later origin, and that there were perfect insects before larvæ and pupæ.

A sequence based on fossil remains would, at present, be unsatisfactory, seeing that at least one Coleopteron (allied to the Curculionidæ) has been found in the Coal-measures, and so far is among the oldest known insects.

Order I. COLLEMBOLA.

Thorax of three, abdomen of six segments; the anterior abdominal segment with a ventral tube or sucker beneath, the penultimate segment with saltatory appendages. Antennæ few-jointed. Wingless. No metamorphosis.

The eyes consist of "distinct ocelli." *Templetonia* has one ocellus on each side, and *Smynthurus* and some others eight. The mouth is not very distinctly mandibulate, and in *Anurida* it is suctorial. The tracheæ are in general well developed. The saltatory appendages consist of two long processes bent under the body and held by a small catch; directly this releases its hold, the spring jerks back, and the creature is thrown upwards and forwards. The tube or sucker contains a viscous fluid. The body is clothed either with hairs or with scales.

These insects are found commonly in damp places and on the surface of pools. *Desoria nivalis* lives on the glaciers of the Alps, the ice being sometimes blackened by its numbers. *Templetonia crystallina* is said to do great mischief to the gutta-percha protecting the underground telegraph-wires of London.

This order was separated from the following, with which it had been included, by Sir J. Lubbock, who, however, does not regard them as true insects in the "strictest sense." They are said to be primitive, not degraded, forms of the original insect stock. They are sometimes associated with the Neuroptera or with the Orthoptera. Gerstaecker places them with the latter as a "tribe" between Thripidæ and the "true" Neuroptera; Claus as a sub-order, also of Orthoptera, before Forficulidæ; Schmarda as an order between Hemiptera and Orthoptera. It is the first order of insects, including Thysanura, according to Von Hayek (1879).

Smynthuridæ.	Templetonia.	Desoria.
Smynthurus.	Orchesella.	
Papirius.		*Lipuridæ.*
		Lipura.
Degeeriidæ.	*Poduridæ.*	
		Anurididæ.
Lepidocyrtus.	Podura.	
Degeeria.	Achorutes.	Anurida.

Order II. THYSANURA.

Thorax of three, abdomen of ten segments; the latter terminated by setiform appendages. Antennæ long, many-jointed. Wingless. No metamorphosis.

The mouth is more distinctly mandibulate. The abdomen beneath is furnished with groups of stiff hairs or with cylindrical appendages. The females have an ovipositor. *Iapyx* and *Campodea* have no eyes; the former has its caudal appendages (cerci) modified into a pair of horny forceps. The nerve-centres are reduced to two ventral ganglia (*Claus*).

Unlike the preceding, these insects are found in dry warm places. They are frequently clothed with hairs or with scales. *Machilis maritima* is common on rocks by the sea. *Campodea* is supposed to be "the representative of a form from which many other groups have been derived." *Lepisma saccharina* is frequently found among sugar.

Campodeidæ.	*Iapygidæ.*	*Lepismidæ.*
Campodea.	Iapyx.	Lepisma.
Nicoletia.		Machilis.

Order III. HEMIPTERA.

RYNGOTA. SIPHONATA. DERMAPTERA.

Four wings, more or less membranous. Mouth produced into an acute suctorial proboscis (mandibulate in Mallophaga). Larva with no trace of wings. Pupa active, semicomplete.

The proboscis is formed by modifications of the labium, mandibles, and maxillæ, and, except in Thripidæ, there are no palpi. The upper wings in the more normal forms are merely coverings (tegmina) for the lower, but they always differ in size and texture; in some cases they are covered by the scutellum. The females have frequently an ovipositor, by which galls are often produced. Ocelli are very commonly present. In some Nepidæ there are caudal setæ connected with anal spiracles and subservient to respiration.

The greater part of the Hemiptera are vegetable feeders, sucking the juices of plants. The Aphides, of which there are about 350 British species, are the most obnoxious; Coccidæ are also very injurious. In the West Indies, *Delphax saccharivora* is very destructive to the sugar-cane. *Coccus cacti*, when dried, is the cochineal of commerce; a pound weight contains 70,000 insects. *Coccus lacca* yields the prepared substance called shell-lac. *Dorthesia* secretes from the end of the abdomen long snow-white flocculent masses of a waxy character. *Ancyra, Flata, Lystra,* and others are also wax-producers; some are so completely enveloped in this secretion as to be unrecognizable. The "Pela," or Chinese wax, is prepared from this substance. Manna is a vegetable secretion caused by *Coccus manniferus*.

The eggs of *Corixa mercenaria* form a food extensively used by the Mexicans. They are collected in freshwater lakes, or are washed on shore. A white limestone rock is forming at the present day in the lakes of Texcono and Chalco from their remains.

There is a tendency to degradation in this order, as shown by the frequent absence of wings, an obsolete mouth, a tarsus reduced to one joint, &c. In the Coccidæ many of the females become more and more inert as they approach the imago state, and the young are sometimes hatched beneath the dead body of the parent.

Westwood places Hemiptera between Diptera and Lepidoptera; Claus (as Rhynchota) between Neuroptera and Diptera.

Besides plant-lice (Aphidæ) and scale-insects (Coccidæ), this order contains the lantern-fly (*Fulgora laternaria*), frog-hopper or cuckoo-spit (*Ptyelus spumarius*), the water boatman (*Notonecta glauca*), the bug (*Cimex lectularius*), and the louse (*Pediculus humanus*). The latter, a very abnormal form, cannot well be separated from the bird-lice (Mallophaga). The number of species in this order cannot be estimated at less than 20,000.

There are five well-marked suborders: the first two are often regarded as forming a distinct order under the name of "Homoptera," while the fourth forms the order "Thysanoptera" of Halliday and Westwood. Mallophaga are not regarded as true insects by the latter; they form the first two of the sixteen families of Neuroptera according to Von Hayek.

Polyctenes, a very remarkable form originally described by Giglioli, and which Westwood considers to have an affinity with the Hemiptera, is now approximated by C. Waterhouse to the Pupipara. It is a parasite on bats; one of its species is from China, the other from Jamaica.

Thorax normal (of three pieces).
  Mouth suctorial: without palpi.
    Wings membranous.
      Tarsi one- or two-jointed .... PHYTOPHTHIRIA.
      Tarsi three-jointed .... HOMOPTERA.
      Anterior wings coriaceous at the base . HETEROPTERA.
    Mouth submandibulate: with palpi ...... THYSANOPTERA.

Thorax abnormal (of two pieces, or three indistinct) .... MALLOPHAGA.

PHYTOPHTHIRIA (= Stenorhynchi).—Tarsi one- or two-jointed. Antennæ of more than six joints. Wings two or four, often wanting.

The pupæ in many cases are not to be distinguished from the imago. The larva, especially in the Aphides, is often covered by a cottony secretion.

Aphides are mostly viviparous in summer, but oviparous in winter; in the former case the females are winged. Parthenogenesis goes on among them through many generations. Whole generations are sometimes resident in gall-like excrescences caused by them. A peculiar secretion (honey-dew) is voided from the anal siphunculi of many species. *Phylloxera vastatrix* is the vine-pest.

Coccidæ have only one-jointed tarsi, the male two-winged, the female apterous.

Coccidæ (Scale-insects).

Coccus.
Lecanium.
Aspidiotes.
Mytilaspis.
Dorthesia.

Aleurodidæ.

Aleurodes.

Aphididæ.

Aphis (Plant-louse).
Lachnus.
Pemphigus.
Chermes.
Tetraneura.
Schizoneura.
Eriosoma.
Adelges.
Phylloxera.

Rhizobius.
Thelaxes.

Psyllidæ.

Psylla.
Livia.
Trioza.
Arytæna.
Aphelara.

HOMOPTERA (=Auchenorhynchi).—Tarsi three-jointed. Antennæ three- or six-jointed. Wings membranous, deflected.

The female is often furnished with an ovipositor composed of a two-valved sheath enclosing a cylindrical horny borer.

Cicadidæ are remarkable for their song; it is confined to the males, and is due to two membranes, acted on by muscles placed in a cavity at the base of the abdomen, and covered externally by the dilated sides of the metasternum.

Iassidæ.

Typhlocyba.
Iassus.
Bythoscopus.
Acocephalus.
Eupelix.

Ledridæ.

Penthimia.
Gypona.
Ledra.

Tettigoniidæ.

Ciccus.
Diatostemma.
Aulacises.
Proconia.
Tettigonia.

Cercopidæ.

Aphrophora.
Ptyelus.

Cercopis.
Eurymela.
Cephalelus.
Ulopa.

Membracidæ.

Centrotus.
Combophora.
Cyphonia.
Bocydium.
Œda.
Heteronota.
Darnis.
Hemiptycha.
Thelia.
Entylia.
Polyglypta.
Aconophora.
Smiliorachis.
Smilia.
Ceresa.
Umbonia.
Pterygia.

Bolbonota.
Enchenopa.
Membracis.

Tettigometridæ.

Tettigometra.

Flatidæ.

Nephesa.
Ricania.
Pœciloptera.
Flata.
Pochazia.

Derbidæ.

Diospolis.
Derbe.
Otiocerus.

Issidæ.

Ancyra.
Eurybrachys.

Hemisphœrius.
Mycterodes.
Issus.

Cixiidæ.

Achilus.
Cixius.
Aræopus.
Asiraca.
Allelophasis.
Delphax.
Phenax.

Lystridæ.

Elidiptera.
Dichoptera.
Dictyophora.
Pœocera.
Aphæna.
Lystra.

Fulgoridæ.

Cyrene.
Enchophora.
Hotinus.

Phrictus.
Pyrops.
Fulgora(Lantern-fly).

Cicadidæ.

Huechys.
Mogannia.
Carineta.
Cicada.
Fidicina.
Dundubia.

———

Cystosoma.

HETEROPTERA.—Tarsi three-jointed. Antennæ with four or five joints. Wings horizontal, the upper pair coriaceous at the base.

The rostrum springs from the fore part of the head, not from beneath as in the Homoptera.

The majority of this group are vegetable feeders, one (*Cimex lectularius*) has man for its prey.

The first three families of the following list form the Hydrocorisæ (=Cryptocerata) of Latreille; the remainder are his Geocorisæ (=Gymnocerata). Hydrocorisæ and the first two families of the Geocorisæ are aquatic. Gerstaecker includes Gerridæ in the Hydrocorisæ; but the latter, with the Hydrometridæ, are the Hydrodromica of Fieber. There are other equally useless divisions and subdivisions and names, whether synonymous or not. In round numbers there may be about 10,000 species in this suborder.

Notonectidæ.

Anisops.
Notonecta.
Ploa.
Sigara.
Corixa.

Nepidæ.

Aphelochira.
Naucoris.
Nepa.
Belostoma.

Ranatra.

Galgulidæ.

Pelogonus.
Galgulus.

Gerridæ.

Hebrus.
Microvelia.
Mesovelia.
Velia.
Gerris.

Hydrometridæ.

Hydrometra = Limnochares.

Saldidæ.

Salda = Acanthia.
Leptopus.

Emesidæ.

Emesodema.
Emesa.
Plœaria.

Reduviidæ.

Nabis.
Prostemma.
Pirates.
Reduvius.
Ectrichodia.
Apiomerus.
Pygolampis.
Harpactor.
Prionotus.
Eulyes.
Sycanus.
Centrocnemis.
Zelus.
Cimbus.

Tingidæ.

Syrtis = Phymata.
Cimex = Acanthia
  (Bug).
Aneurus.
Aradus.
Agramma.
Monanthia.
Dictyonota.
Orthostira.
Derephysia.
Zosmenus.
Tingis.

Capsidæ.

Bryocoris.
Halticus.
Pithanus.
Calocoris.
Phytocoris.
Psallus.
Heterotoma.
Lygus.
Miris.

Capsus.

Lygæidæ.

Microphysa.
Geocoris = Ophthal-
  micus.
Henestaris.
Pyrrhocoris.
Anthocoris.
Ischnodemus.
Plinthisus.
Rhyparochromus.
Orsillus.
Nysius.
Phygadicus.
Lygæus.

Berytidæ.

Metacanthus.
Metatropis.
Berytus.
Neides.

Coreidæ.

Micrelytra.
Chorosoma.
Alydus.
Phyllomorpha.
Verlusia.
Coreus = Dasycoris.
Syromastes.
Atractus.
Rhopalus.
Stenocephalus.

Pentatomidæ.

Edessa.
Phlœa.
Ælia.
Eusarcoris.

Mormidea.
Atelocera.
Acanthosoma.
Rhaphigaster = Ne-
  zara.
Palomenus.
Pentatoma.
Strachia.
Sciocoris.
Sehirus.
Cydnus.
Æthus.
Zicrona.
Arma.
Picromerus =
  Asopus.
Canthecona.
Stiretrus.

Scutelleridæ.

Coptosoma = Thyreo-
  coris.
Probænops.
Plataspis.
Corimelæna.
Podops.
Phimodera.
Ancyrosoma.
Psacasta.
Tetyra = Eurygaster.
Symphylus.
Agonosoma.
Graphisoma.
Trigonosoma.
Odontotarsus.
Pachycoris.
Callidea.
Scutellera.
Cantao.
Sphærocoris.
Tectocoris.

THYSANOPTERA (=Physopoda).—Mouth with four palpi, the mandibles forming a short rostrum. Wings nerveless, fringed.

The tarsi are two-jointed, the last joint, instead of claws, having a sort of sucking-disk. In some species the female is provided with an ovipositor.

A limited group of small insects living under the bark of trees, or in flowers, and sucking their sap. Claus places them in the Pseudo-Neuroptera.

Thripidæ.
Coleothrips.
Thrips.

MALLOPHAGA (= Anoplura).—Meso- and metathorax united. Mouth mandibulate. Wingless. No metamorphosis. Parasitic.

The mouth has two mandibles, with or without palpi. The antennæ are from three- to six-jointed. The eyes are simple, but sometimes absent. The thorax is apparently composed of two rings.

The Mallophaga are Hemiptera, which, owing to their parasitic nature, have not passed beyond a primary stage of development. They are found mostly on birds, living on the feathers. The Pediculidæ, which must be placed near them, have the mouth produced into a fleshy proboscis, with hooks at the end, within which, enclosed in a chitinous sheath, are two sharp-pointed stylets. They live on the blood of mammals. *Hæmatomyzus* is a peculiar parasite of the Indian elephant.

Nirmidæ.	Docophorus.	Gyropus.
Trichodectes.		Menopon.
Nirmus.	*Liotheidæ.*	Eureum.
Goniodes.	Liotheum.	Colpocephalum.
	Trinotum.	

Pediculidæ.		Hæmatopinus.
Pediculus (Louse).		
Phthirius. ·		Hæmatomyzus.

Order IV. ORTHOPTERA.

ULONATA.

Four wings, two anterior coriaceous, pergameneous, or submembranous; two posterior membranous, folded longitudinally. Mandibulate. Larva and pupa more or less resembling the imago, but without wings.

The upper wings are variable in texture, either meeting at the edges or overlapping more or less, but they are never exactly like the lower pair. Many species are apterous, especially among the females. Ocelli are frequently present. Some have an exserted ovipositor. Earwigs (Forficulidæ) are armed with a caudal forceps, and the crickets (Gryllidæ), Mantidæ, &c. have setaceous appendages (cerci), which are common to both sexes, and are supposed to be connected with the sense of touch. Besides these ordinary anal appendages, there is in some one or two pairs of stylets, occasionally, in the male, modified into hooks.

In their internal anatomy the Orthoptera are very highly organized, and are considered to stand in the foremost ranks of insects. None of them are aquatic in any stage of their existence. *Prisopus flabellicornis* (Brazilian) is said, however, to spend the day under water, attached to stones &c., and to fly about only at dusk.

The noise produced by male crickets and Locustidæ (the females are silent) is due to the anterior wings rubbing against each other at a part where both are furnished with a rasp-like nervure; while in grasshoppers and locusts (Acridiidæ) it is the posterior femora rubbing against the anterior wings. In the former the males are also provided with a talc-like spot at the base of the anterior wings.

In the cockroaches (Blattidæ) the eggs are deposited in a capsule, which is carried by the female, and in which the larvæ are hatched. The Mantidæ deposit their eggs in cases of a papery texture divided into cells, which they make under stones or on the twigs of plants. The Saltatoria generally place their eggs in the earth, often forming clay-tubes for their reception.

Except the Mantidæ and omnivorous Blattidæ, all are vegetable feeders, and, being very prolific, frequently destroy the produce of a whole district. The locust (*Œdipoda migratoria*) is only too well known. Earwigs are great enemies of bees, as well as being herbivorous.

The only special auditory organs in insects are found in Acridiidæ (a small membranous expansion on each side of the metathorax) and in Locustidæ and Gryllidæ (a similar membrane on each of the fore tibiæ).

The cockroach (*Periplaneta orientalis*), mole cricket (*Gryllotalpa vulgaris*), cricket (*Gryllus domesticus*), locust (*Œdipoda migratoria*), [the North-American locust is *Caloptenus spretus*], and earwig (*Forficula auricularia*) belong to this order. Grasshoppers are of various species, and belong to *Stenobothrus* and *Œdipoda*, but the latter genus is not found in England. The stick- and leaf-insects are Phasmidæ; but the anterior wings, resembling leaves, are also found in certain Locustidæ.

I

It is very unfortunate that locusts should not have been retained under the old Latin name of *Locusta*; but the nomenclature here given is now universally adopted.

Thysanura and Collembola have been placed in this order.

With it also have been combined Dragonflies and other Neuroptera, the so-called incomplete metamorphosis being the only character in common. Perhaps it would be more convenient to unite the two; naturally they appear to include ten distinct groups.

Of the four suborders of Orthoptera, Euplexoptera [=Dermaptera] is so far differentiated that Westwood considers it entitled to ordinal rank. Blattidæ has also been elevated into an order [Dictyoptera] by Leach. They are both included in the Cursoria by Claus.

> Anal segment without a forceps.
>> Hind legs formed for walking.
>>> Head retracted .... CURSORIA.
>>> Head exserted .... GRESSORIA.
>> Hind legs formed for leaping .... SALTATORIA.
> Anal segment with a movable forceps ... EUPLEXOPTERA.

CURSORIA.—Body ovate, depressed. Head retracted into the prothorax. Legs slender; tarsi with an accessory joint [plantula] between the claws.

The antennæ are generally very long and slender. The upper wings are coriaceous, the lower longitudinally folded; but they are not often developed in the female, and not always in the male.

These insects are mostly nocturnal, and are vegetable feeders, but many are omnivorous.

Blattidæ.	Blatta.	Gynopeltis.
Polyzosteria.	Thyrsocera.	Euthyrhapha.
Perisphæria.	Periplaneta (Cock-	Panesthia.
Heterogamia.	roach).	Hormetica.
Corydia.	Panchlora.	Blabera.
Phoraspis.	Derocalymma.	Monachoda.

GRESSORIA.—Body long, narrow. Head exserted. Legs slender; posterior femora not thickened.

In the Mantidæ the anterior legs are raptorial and their femora grooved beneath for the reception of the tibiæ in repose. The prothorax is by far the longest of the thoracic segments; in the Phasmidæ it is the shortest. Both families are remarkable in that some of their species have the upper wings resembling leaves, hence "walking leaves." In those Phasmidæ devoid of wings, the resemblance is to a twig or stick, "walking sticks." Although very often sluggish in their movements, some are capable of taking short leaps.

The Mantidæ are very voracious animal feeders: the Phasmidæ live entirely on vegetable matter; they are amongst the longest of insects—*Bacteria sarmentosa* is ten inches in length.

Mantidæ.
Eremiaphila.
Schizocephala.
Acanthops.
Toxoderes.
Harpax.
Empusa.
Ameles.
Mantis.
Blepharis.
Pyrgomantis.

Metallyticus.
Tarachodes.
Deroplatys.
Hoplophora.

Phasmidæ.
Bacillus.
Bacteria.
Lonchodes.
Cladoxerus.

Cyphocrania.
Podacanthus.
Diapherodes.
Euracantha.
Phasma.
Necroscia.
Aschiphasma.
Prisopus.
Metriotes.
Extatosoma.
Phyllium.

SALTATORIA.—Body generally slender. Hind legs saltatorial, the femora thickened.

These are voracious vegetable feeders, some of the species occurring in vast numbers at uncertain intervals.

In Gryllidæ and Locustidæ the antennæ are long and setaceous, with sometimes as many as 140 joints; in Acridiidæ they are short and stout, rarely clavate. The ovipositor in the two former families is often very large; in Acridiidæ there is no ovipositor. *Hemimerus* is an aberrant form allied to *Gryllotalpa* according to Walker, but for De Saussure forming a distinct order (Diploglossata).

Gryllidæ.
Gryllotalpa (Mole-cricket).
Cylindrodes.
Xya=Tricondyla.
Myrmecophila=Sphærium.
Gryllus=Acheta (Cricket).
Œcanthus.
Nemobius.

Platyblemma.
——
Hemimerus.

Locustidæ.
Saga.
Phalangopsis.
Locusta.
Decticus.
Phyllophora.
Pterochroza.

Cycloptera.
Cyrtophyllum.
Phaneroptera.
Meconema.
Barbitistes=Odontura.
Conocephalus.
Xiphidium.
Hetrodes=Eugaster.
Anostostoma=Dinacrida.

Ephippigera.
Schizodactylus.

Acridiidæ.

Proscopia.
Tryxalis.
Œdipoda = Pachy-
tylus (Locust).

Stenobothrus.
Gomphocerus.
Opomala.
Acridium.
Caloptenus.
Romalea.
Choriphyllum.

Hymenotes.
Chorotypus.
Pamphagus = Por-
thetis.
Pneumora.
Tettix = Tetrix.
Ommexecha.

EUPLEXOPTERA.—Anal segment provided with a movable forceps. Under wings folded both transversely and longitudinally. No ocelli.

The upper wings are very short, coriaceous, and without veins. The antennæ are filiform, with from twelve to forty joints.

These are chiefly vegetable feeders, but some are carnivorous. They are mostly nocturnal insects.

Forficulidæ (Earwigs).

Labia.
Labidura.
Pygidicrania.

Forficula.
Brachylabis.
Apachya.

Order V. NEUROPTERA.

Four more or less equal membranous wings, generally reticulated, rarely folded. Mandibulate. Pupa incomplete, more or less resembling the imago, either quiescent or active. Larva with six articulated legs.

There are exceptions to all the leading characters of this order, especially in the transformations, which are often of an anomalous character. In the imago state the wings are sometimes wanting, or only present in the male; in Perlidæ they are longitudinally folded. In Ephemeridæ there are no mandibles. One family, Termitidæ, contains neuters of two kinds—soldiers and workers—which, with the females, but only after pairing, are wingless.

It will thus be seen that the order comprises several discordant groups, dissimilar in their leading characters as well as in their habits; and their earlier states are equally diversified. The four nearly equal wings are said best to distinguish them.

Fabricius (Syst. Ent. 1775), in his purely artificial system, classed spiders and centipedes with dragonflies in his order Unogata; a few years later (*Mantissa*, 1787) he proposed Synistata for the remainder of the Neuroptera, including also the Hymenoptera, as well as *Oniscus, Monoculus*, &c.

The Planipennia alone constitute the Neuroptera according to the views of some authors, the remainder, including Trichoptera, being placed with the Orthoptera, under the name of "Pseudo-Neuroptera" (Archiptera of Häckel).

According to Packard, Libellulidæ and the Ephemeridæ are the most typical of the Neuroptera. These families and the Perlidæ are among the earliest known insects. They seem to occur most abundantly in the Devonian formations of North America.

To this order belong the destructive white ants, *Termes bellicosus* and other species; the death-watch, *Atropos pulsatoria*; the may-fly, *Ephemera vulgata*; and ant-lion, *Myrmeleon formicaleo*. Libellulidæ are known generally as dragonflies or horse-stingers, and in the United States as the "Devil's darning-needles." It is curious to contrast these names with that of the French, "Demoiselles." *Chrysopa*, of which there are many species, is known as the "Golden-eye."

The earlier stages of these insects are very variable. In the Libellulidæ, Termitidæ, and Ephemeridæ, although very unlike each other, the metamorphosis is semicomplete, while *Chloëon* undergoes a series of not less than twenty moultings, in some stages a retrograde movement taking place. Prof. Westwood, with regard to the earlier states of the order, proposes two primary divisions:—" 1. Those with an active pupa, undergoing a metamorphosis which, for want of a better name, we may, with MacLeay, term sub-semicomplete; in all which there is a greater dissimilarity between the larva and imago states than exists in the insects typical of the monomorphous, semicomplete metamorphosis. Here belong the Psocidæ and Termitidæ, which have terrestrial larvæ, and the Libellulidæ, Ephemeridæ, and Perlidæ, which are aquatic in their preparatory states. I term the species of this division Biomorphotic insects. 2. Those which have quiescent incomplete pupæ, which, however, acquire the power of locomotion shortly before the assumption of the perfect state. This division (Subnecromorphotica) comprises the families Myrmeleonidæ, Hemerobiidæ, Sialidæ, Panorpidæ, Raphidiidæ, and Mantispidæ." The aquatic larvæ breathe by means of filamentous branchiæ.

Retaining, with M'Lachlan, the order Neuroptera "as a matter of convenience only," it may be naturally divided into seven suborders:—

Wings naked, rarely a few hairs or scales.
Wings folded in repose .... PLECOPTERA.

Wings not folded.
   Jaws well developed.
      With wingless neuters .... ISOPTERA.
      No neuters.
         Antennæ short, few-jointed............... ODONATA.
         Antennæ longer, many-jointed
            Wings with few veins .... CORRODENTIA.
            Wings with many veins .... PLANIPENNIA.
   Jaws rudimentary or obsolete.................. AGNATHI.
Wings hairy .... TRICHOPTERA.

CORRODENTIA.—Antennæ many-jointed. Wings few-veined. Tarsi 2- to 3-jointed.

The posterior wings of Psocidæ are small or rudimentary, or both pairs may be wanting. They live on dead animal and vegetable matter, and are mostly arboreal in their habits. *Atropos pulsatoria* is the pest of insect collections. Embiidæ have moniliform antennæ. They are vegetable feeders, and construct silken galleries under stones.

Psocidæ.	*Embiidæ.*
Psocus.	Embia.
Atropos = Troctes.	Olyntha.
Cæcilius.	Oligotoma.

ISOPTERA.—Antennæ rather short, many-jointed. Wings very large, equal. Neuters.

The larvæ and pupæ resemble the neuters. These insects are social in their habits, building large nests on trees and on the ground, and are very destructive to woodwork. The female of *Termes bellicosus* is said to lay 80,000 eggs in twenty-four hours.

Termitidæ.
Termes (White Ant).
Calotermes.

PLECOPTERA.—Antennæ long, setaceous, many-jointed. Wings subsimilar, folded longitudinally in repose. Jaws rudimentary.

The larvæ and pupæ are aquatic; the latter, except in wanting wings, resemble the imago. Some species have caudal setæ.

Perlidæ.	
Nemura.	Isopteryx.
Capnia.	Perla.
	Pteronarcys.

AGNATHI.—Antennæ short, setaceous, of three joints, the last very long. Jaws soft, membranous or obsolete. Posterior wings small, sometimes wanting. Abdomen ending in two or three long setæ.

The fore legs are long in the male. The larvæ are aquatic, live two or three years; the perfect insect takes no food, and dies in a day or two. They sometimes appear in almost fabulous numbers, and always in the evening.

Ephemeridæ.	Chloëon.	Palingenia.
Oligoneura.	Baetis.	Ephemera (May-fly).
Lachlania.	Heptagenia.	Oniscigaster.
Cænia.	Potamanthus.	*Platephemera.

*Hemeristia. *Breyeria.

ODONATA.—Antennæ short, setaceous, of seven joints at most, the last subulate. Jaws very strong, covered by the lips. Wings subequal, reticulate.

The abdomen in the male ends in two curved, in the female in two simple appendages. The eyes are very large. The larva and pupa are aquatic, the former very dissimilar to the perfect insect; in all stages they are voracious insect feeders.

The three groups given below are not strictly entitled to the rank of families. They are all known as "dragonflies."

Agrionidæ.	*Libellulidæ.*	*Æschnidæ.*
Agrion.	Cordulia.	Æschna.
Lestes.	Epitheca.	Anax.
Callepteryx.	Libellula.	Gomphus.
Platycnemis.	Diplax.	Cordulegaster.

PLANIPENNIA.—Antennæ elongate, many-jointed. Jaws distinct. Wings nearly equal.

The larvæ mostly terrestrial; in Sialidæ they are aquatic, but sometimes they live out of the water before transformation. Pupæ incomplete, inactive. Myrmeleontidæ and Ascalaphidæ have clavate antennæ. Coniopterygidæ are covered with a whitish powder. In Rhaphidiidæ the female has an ovipositor. Panorpidæ have a clypeus lengthened into a snout with very small mandibles at the end. The perfect insects are generally herbivorous, but the larvæ are voracious insect feeders.

Myrmeleontidæ.
Myrmeleon (Ant-lion).

Ascalaphidæ.
Stilbopteryx.
Pamexis.
Ascalaphus.
Palpares.

Hemerobiidæ.
Sisyra.
Micromus.
Hemerobius.

Psectra.
Psychopsis.
Rapisma.
Chrysopa.
Osmylus.
Nemoptera.
Nymphes.

Coniopterygidæ.
Coniopteryx.

Mantispidæ.
Mantispa.
Trichoscelia.

Rhaphidiidæ.
Rhaphidia.

Sialidæ.
Chauliodes.
Corydalis.
Sialis.

Panorpidæ.
Boreus.
Merope.
Bittacus.
Panorpa (Scorpion-fly).

TRICHOPTERA (Caddis-flies).—Four membranous wings; the anterior generally hairy, the posterior folded when at rest. Mandibles rudimentary.

The antennæ are many-jointed, setaceous. The maxillary palpi vary considerably in structure. The wings have longitudinal nervures and few transverse ones; and in repose they are closed up and deflexed in an almost vertical manner. The hairs covering them are mostly simple, but occasionally inflated and scale-like. The abdomen has nine segments, the last bearing appendages of very varied kinds. The tibiæ are furnished with fine spines and also with movable spurs; the tarsi have a pair of simple claws, between which is a short cushion [plantula] and two membranous "arolia."

The larvæ are six-footed and aquatic, and construct cases of bits of leaves, twigs, straw, sand, or shells; in these cases the pupa stage is assumed.

M'Lachlan gives eight families for the European fauna. These families include all the species known at present.

Phryganeidæ.
Phryganea.
Neuronia.
Agrypnia.

Limnophilidæ.
Glyphotælius.
Limnophilus.

Stenophylax.
Halesus.
Chætopteryx.
Thamastes.
Enoicyla.

Sericostomatidæ.
Sericostoma.

Goëra.
Silo.
Nosopus.

Leptoceridæ.
Setodes.
Leptocerus.
Mystacides.
Triænodes.

Œstropsidæ.	Lype.	Agapetus.
Œstropsis.	Psychomyia.	Beræa.
	Polyceutropus.	

Hydropsychidæ.   *Hydroptilidæ.*

Hydropsyche.   *Rhyacophilidæ.*   Agraylea.

Diplectrona.   Chimarrha.   Hydroptila.

Tinodes.   Rhyacophila.

Order VI. DIPTERA.

ANTLIATA.

Wings two, membranous, never folded, with radiate nervures; posterior wings replaced by a small clavate filament on each side [halteres]. A suctorial proboscis. Prothorax very short. Metamorphosis complete. Larva apodal.

In this order the proboscis is composed of mandibles, maxillæ, and a central piece or tongue [glossarium], the labium, often with a fleshy terminal lip, serving as a sheath, and they are often converted into chitinous setæ or into lancet-shaped bodies. There are commonly two maxillary, but no labial palpi. The antennæ are in general very short, of three joints, the last furnished with a bristle; in the Nemocera the antennæ are jointed in the ordinary manner. The eyes are large, and there are three ocelli. The pro- and metathorax are much reduced, and the mesothorax correspondingly enlarged. The wings have often a membranous appendage at the base [alula]; there is also frequently a small scarious scale [squamula] placed behind the insertion of the wings. The halteres are attached to the metathorax; they have been supposed to be subservient to respiration, to hearing, and to smell. The abdomen in the female is sometimes terminated by a sort of telescopic ovipositor. The foot, besides its two claws, is provided with two, rarely three, cushions [pulvilli], covered beneath with fine hairs, expanded at their tips, supposed to act as suckers, probably aided by the secretion of a viscid matter.

Many of the Diptera are useful scavengers in the larval state, but others are very injurious—e. g. *Cecidomyia destructor* to wheat-crops, *Tipula oleracea* to grass-lands, *Dacus oleæ* to the fruit of the olive, &c. In the perfect state they are too often pests to man and beast, sucking the blood, or depositing their eggs in or on their bodies, causing tumours, ulcerations, and death.

The fecundity and rapid succession of generations of many species cause them to appear sometimes in immense quantities; the progeny of the blowfly is said to amount to 500 millions in twelve months. Agamic reproduction occurs in some species; the larva of *Miastor metraloas* has been found to contain other larvæ identical in all respects, except in size, with the enveloping larva; and these larvæ continue to produce successive generations which ultimately develop into perfect insects.

In the Pupipara the larva changes into the pupa in the abdomen of the mother; in all other Diptera eggs, or larvæ just released from the egg, are produced. The larvæ, known as maggots, are apodal, or very rarely with rudimentary legs. In the blowfly " not one structure exists as it exists in the maggot " (*Lowne*). The mouth is generally provided with two hooks, which "are thrust into the substance from which the larvæ derive their nutriment." The pupa, in the majority of Diptera, is merely the larva with a hardened case [pupæ coarctatæ]; in the remainder the pupa is in a separate case, having its limbs enclosed in distinct sheaths [pupæ obtectæ]. The pupæ of the Culicidæ are locomotive.

The species are very numerous; about 9000 are found in Europe; Schiner thinks that they are not a twentieth part of those inhabiting the whole world. Among them are the gnat (*Culex pipiens*), musquito (*Culex*, sp. variæ), Hessian fly (*Cecidomyia destructor*), gad-fly (*Tabanus bovinus*), fly (*Musca domestica*), blue-bottle (*Musca Cæsar*), blow-fly (*Musca vomitoria*), flesh-fly (*Sarcophaga carnaria*), tzetze (*Glossina morsitans*), cheese-hopper (*Piophila casei*), bot-fly of the sheep (*Œstrus ovis*), bot-fly of the horse (*Gastrus equi*), bot-fly of the bullock (*Hypoderma bovis*), spider-fly, forest-fly or bot (*Hippobosca equina*), sheep-tick (*Melophagus ovinus*), and flea (*Pulex irritans*). Sand-flies are a general name; some are referable to *Simulium*.

The Diptera are divided into four sections or suborders; but the last is aberrant, and is by some ranked as an order. Brauer has divided them according as the pupa-case opens longitudinally (Orthorhapha) or curvilinearly (Cyclorhapha).

Thorax distinct from the abdomen.
  Larva developed from the egg.
    Antennæ many-jointed...................... NEMOCERA.
    Antennæ three-jointed...................... BRACHYCERA.
  Larva and pupa developed in the body of the
    mother .... PUPIPARA.
Thorax confounded with the abdomen .... APHANIPTERA.

PUPIPARA.—Larva and pupa developed in the abdomen of the mother. Head retracted. Antennæ in a cavity of the head. Parasitic.

The mouth is peculiar and the analogies of its parts very obscure. The wings are often rudimentary or absent, and with or without halteres. Eyes and ocelli are sometimes wanting. The bee-louse (*Braula cæca*) is a parasite on the Italian bee (*Apis Ligustica*). *Nycteribia* is a spider-like wingless form found on bats.

Braulidæ.

Braula (Bee-louse).

Nycteribiidæ.

Nycteribia.

Hippoboscidæ.

Hippobosca (Bot, or Forest-fly).
Melophagus (Sheep-tick).
Ornithomyia.
Strebla.

BRACHYCERA.—Oviparous. Antennæ short, apparently not more than three-jointed. Palpi one- or two-jointed.

The larvæ are aquatic or terrestrial, feeding on vegetable or animal matter, or parasitic. The perfect insect lives on the juices of animals or plants.

The antennæ are many-jointed in some families (Beridæ, Stratiomyidæ, Tabanidæ), the so-called third joint really consisting of several. The number of pieces composing the haustellum varies—two, four, or six; and on this character Macquart has founded his arrangement, naming his divisions Dichætæ, Tetrachætæ, and Hexachætæ respectively. Tabanidæ, in having nearly all the parts of a mandibulate mouth, should probably be considered as the highest form; and the parasitic Œstridæ, which have an obsolete mouth, the lowest. It is perhaps, however, more convenient to adopt the sequence of Macquart.

HEXACHÆTÆ.

Tabanidæ.

Hæmatopota.
Chrysops.
Tabanus (Gad-fly).
Pangonia.

TETRACHÆTÆ.

Beridæ.

Xylophagus.
Actina.
Beris.

Stratiomyidæ.

Cœnomyia.
Clitellaria.
Sargus.

Pachygaster = Vappo.
Nemotelus.
Platyna.
Odontomyia.

Stratiomys.
Oxycera.

Mididæ.

Midas.
Apiocera.

Asilidæ.

Leptogaster.
Dioctria.
Dasypogon.
Mallophora.
Laphria.
Asilus.

Hybidæ.

Ocydromia.
Hybos.

Empidæ.

Tachydromia.
Platypalpus.
Hilara.

Ramphomyia.
Empis.

Acroceridæ.

Panops.
Oncodes = Henops.
Acrocera.

Bombyliidæ.

Lomatia.
Anthrax.
Bombylius.
Nemestrina.

Leptidæ.

Atherix.
Leptis.

Therevidæ.

Thereva.

Dolichopodidæ.

Argyra.

Chrysotus.
Medeterus.
Dolichopus.
Rhaphium.

Syrphidæ.

Ceria.
Chrysotoxum.
Paragus.
Pipiza.
Chrysogaster.
Syritta.
Eumerus.
Xylota.
Milesia.
Merodon.
Eristalis.
Volucella.
Rhingia.
Chilosia.
Doros.
Syrphus.
Ascia.
Baccha.

DICHÆTÆ.

Scenopinidæ.

Scenopinus.

Pipunculidæ.

Pipunculus.

Lonchopteridæ.

Lonchoptera.

Platypezidæ.

Platypeza.

Conopidæ.

Myopa.
Zodion.

Stachynia.
Conops.

Œstridæ.

Hypoderma.
Œstrus (Bot-fly).
Gastrus.

Muscidæ.

Echinomyia.
Gonia.
Nemoræa.
Eurygaster.
Metopia.
Tachina.
Exorista.

Miltogramma.
Lophomyia.
Ocyptera.
Gymnosoma.
Phasia.
Dexia.
Sarcophaga.
Stomoxys.
Achias.
Glossina.
Musca (Fly).
Mesembrina.
Aricia.
Hydrophoria.
Sepedon.
Anthomyia.

Cœnosia.
Tetanocera.
Loxocera.
Chyliza.
Cordylura.
Scatophaga.
Sapromyza.
Sciomyza.
Dorycera.
Ortalis.
Dacus.
Ceratites.
Platystoma.

Urophora.
Tephritis.
Trypeta.
Sepsis.
Nemopoda.
Diopsis.
Elaphomyia.
Calobata.
Micropeza.
Calopa.
Lauxania.
Celyphus.
Ochthera.

Cænia.
Ephydra.
Piophila.
Borborus.
Drosophila.
Oscinis.
Chlorops.
Meromyza.
Agromyza.
Phytomyza.

Phoridæ.

Phora.

NEMOCERA.—Antennæ with 6–16 joints. Palpi 4–5-jointed. Larvæ frequently aquatic and free-swimmers; variously feeding according to the family.

The antennæ are often plumose, especially in the males. Culicidæ have a long proboscis of seven pieces. In Tipulidæ and the other families the proboscis is short and the number of pieces vary. Cecidomyiidæ are gall-makers. Psychodidæ are small moth-like flies, with hairy wings and body.

Bibionidæ.
Scatopse.
Simulium.
Aspistes.
Bibio.

———

Chionea.

Tipulidæ.
Limnobia.
Geranomyia.
Tipula (Daddy-long-legs).

Ptychoptera.
Ctenophora.
Dixa.

Mycetophilidæ.
Rhyphus.
Platyura.
Molobrus=Sciara.
Mycetophila.

Cecidomyiidæ.
Cecidomyia (Hessian-fly).

Psychodidæ.
Psychoda.

Chironomidæ.
Ceratopogon.
Corethra.
Chironomus.
Tanypus.

Culicidæ.
Miastor.
Anopheles.
Culex (Gnat).

APHANIPTERA.—Thorax not distinctly marked off from the abdomen; the former with two scales on each side [abortive wings]. Larva vermiform, with a distinct head and jaws. Pupa inactive. Perfect insect an animal-sucker. An aberrant group.

The sucking-apparatus is made up of two elongated mandibles, a sharp slender tongue (labrum), and a sheath formed by the labial palpi. There are also two long maxillary palpi. The antennæ are very small, placed in a cavity, and with the number of joints varying according to the species. The hind legs are formed for leaping.

Pulicidæ.

Pulex (Flea).

Sarcopsylla (Chigoe, or Jigger).

Order VII. LEPIDOPTERA.

Glossata.

Four extended wings, scaly on both sides; nervures branching. Tibiæ spurred. Haustellate or antliate. Larva or caterpillar mandibulate, with six legs and four to ten prolegs. Pupa obtected.

The suctorial proboscis [antlia], or tongue, as it is incorrectly called, consists of two slender pieces [elongated maxillæ], and is rolled up in a state of repose; it is accompanied by two well-developed labial palpi. The head is attached to the thorax by slender ligaments. The body is hairy. The legs are long, slender, and similar, and very loosely attached. In some butterflies the fore legs are rudimentary. A pair of scales [tegulæ, parapter, or pterygoda] are found on the mesothorax at the base of the upper or anterior wings; and a pair of vesicular bodies [patagia or tippets] close to these, but attached to the prothorax. There are also peculiar plumose scales on the wings of the males of certain butterflies. Some of the moths have a special apparatus for keeping the wings on the same side together during flight. The scales covering the wings are very variable, even on different parts of the same insect.

The species of many genera of butterflies and moths closely resemble species of other genera, *Trochilium* and *Sesia* simulating bees, wasps, and flies, and some even beetles. One (*Callima Inachis*) might be taken for a leaf when at rest. Dimorphism or polymorphism is probably not uncommon. *Papilio Pammon* has three sets of females, *P. Memnon* and some others two. Butler remarks that species "almost identical in every respect" in their perfect state are in their larval condition so dissimilar as to leave no doubt of their being distinct; Bates, indeed, considers the value of the larval structure as a systematic character to be "very small."

The larvæ, or caterpillars, before passing into the pupa stage, form cocoons of silk, or silk mixed with other substances, such as bits of wood or their own hairs. The spinneret is a modification of the larval lip. The silk-glands are two long lateral sacs. Caterpillars. are almost invariably vegetable feeders, rarely of flowers, and are often very destructive; a few live in the water. No species is known to be parasitic in any stage of its existence. *Galleria mellionella* and *Achræa grisella*, although found in bee-hives, feed only on the wax.

The affinities of the Lepidoptera are not very decided; some of the Neuropterous Trichoptera have an outside resemblance, and so have species of Hemiptera, Diptera, and Hymenoptera. The species are very numerous, but the colours, on which so many so-called species depend, are very variable. The nomenclature, too, in the attempt to restore obsolete (and in some cases mere catalogue) names, is not uniform. Indeed in this respect the generic names are so unstable that many amateurs are content to omit them altogether. Among the moths the names given to species are often simply absurd, as has been pointed out by Stau-dinger in such words as "*Schmidtiiformis*" or "*Millieridactylus*"; and the silly custom of requiring a uniform termination to the specific names of certain groups, as "*-ella,*" "*-aria,*" "*-ata,*" "*-ana,*" "*-alis,*" &c., can now only mislead, as they are no longer strictly confined to such groups, to which, when *Phalæna* included almost all known moths, they were supposed to afford a kind of clue.

A curious set of English names are given by collectors to butter-flies and moths, especially the latter; but the following names in common use may be cited as species of this order:—The white cabbage butterfly (*Pieris brassicæ*); the peacock (*Vanessa Io*); the brimstone (*Gonopteryx rhamni*); the painted lady (*Pyrameis cardui*) [a cosmopolitan species]; Atlas moth (*Attacus Atlas*); the silkworm moth (*Bombyx mori*); the death's-head (*Acherontia atropos*); the humming-bird moth (*Macroglossa stellatarum*); the goat-moth (*Cossus ligniperda*); the puss-moth (*Cerura vinula*); the magpie moth (*Abraxas grossulariata*); the dart-moth (*Noctua segetum*, very injurious in the caterpillar state to turnip-crops); the clothes-moths are various species of Tineidæ. The larvæ of Geometridæ are well known under the name of "loopers;" those of the Tineidæ mostly feed on the parenchyma of leaves and are known as "leaf-miners."

HETEROCERA (Moths).—Antennæ pectinate, setaceous, plumose or fusiform, rarely clavate. Frequently nocturnal.

In the males the antennæ are often more developed than in the females. The first four families are known as the "Micro-lepidoptera". Sesiidæ are of uncertain affinity.

Pterophoridæ.
Alucita.
Pterophorus.

Tineidæ.
Butalis.
Chrysoclista.
Laverna.
Elachista.
Bucculatrix.
Cemiostoma.
Nepticula.
Coleophora.
Gelechia.
Depressaria.
Lithocolletis.
Cleophora.
Hyponomeuta.
Adela.
Tinea.

Tortricidæ.
Peronæa.
Eupœcilia.
Grapholitha.
Carpocapsa.
Lozotænia.
Penthina.
Conchylis.
Teras.
Tortrix.

Pyralididæ.
Achrœa.
Galleria.
Botys.
Myelois.
Acentropus.
Cataclysta.
Eudorea.
Pempelia.
Crambus.
Asopia.
Pyralis.

Geometridæ.
Psodos.
Cidaria.
Nyssia.
Biston.
Hybernia.
Gnophos.
Boarmia.
Cleora.
Zerene.
Ennomos.
Anaitis.
Melanippe.
Emmelesia.
Eupithecia.
Abraxas.
Fidonia.
Selenia.
Urapteryx.
Rumia.
Acidalia.
Geometra.

Uraniidæ.
Thaliura.
Urania.

Noctuidæ.
Hypena.
Brepha.
Mormo = Mania.
Erastria.
Acontia.
Heliothis.
Anarta.
Nonagria.
Cucullia.
Leucania.
Xanthia.
Cosmia.
Thyatira.
Diphthera.
Catocala.
Plusia.

Aplecta.
Hadena.
Dianthœcia.
Polia.
Miselia.
Acronycta.
Diloba.
Apamæa.
Mamestra.
Charæas.
Xylophasia.
Caradrina.
Orthosia.
Agrotis.
Triphæna.
Noctua.

Psychidæ.
Fumea.
Œceticus.
Psyche.

Lithosiidæ.
Gnophria.
Lithosia.
Setina.
Calligenia.
Nola.
Nudaria.

Arctiidæ.
Notodonta.
Limacodes.
Cerura = Dicranura.
Orgyia.
Stauropus.
Pygæra.
Porthesia.
Liparis.
Spilosoma = Phrag-
    matobia.
Chelonia.
Arctia.
Euthemonia.
Callimorpha.

Euchelia.
Deiopeia.
Emydia.

Bombycidæ.

Clisiocampa.
Lasiocampa = Gas-
  tropacha.
Endromis.
Odonestis.
Attacus.
Saturnia.
Bombyx.

Zygænidæ.

Dioptis.

Ino = Procris.
Cydosia.
Syntomis.
Zygæna = Anthrocera.

Agaristidæ.

Eusemia.
Agarista.
Hecatesia.

Castniidæ.

Castnia.
Synemon.

Hepialidæ.

Cossus.

Zeuzera.
Macrogaster.
Hepialus.

Sphingidæ.

Acherontia.
Deilephila.
Chærocampa.
Smerinthus.
Ambulyx.
Macroglossa.
Sphinx.

Sesiidæ.

Trochilium.
Sesia = Egeria.

RHOPALOCERA (Butterflies).—Antennæ terminated by a club or knob. Diurnal.

A few exceptions occur in which the antennæ are filiform or even pectinated; in the Hesperidæ they are hooked at the tip. Lycænidæ, Erycinidæ, and Nymphalidæ have small or rudimentary fore legs; in the two former this peculiarity is confined to the males.

Hesperiidæ(Skippers).

Thanaos.
Erinnys.
Thymele = Syrich-
  thus.
Cyclopides.
Pyrgus.
Phareas.
Pyrrhopyga.
Pamphila.
Hesperia.

Lycænidæ.

Amblypodia.
Dipsas.
Thecla.
Loxura.
Polyommatus.
Lycæna.

Erycinidæ.

Lampides.

Darnis.
Theope.
Nymphidium.
Lemonias.
Mesosemia.
Mesene.
Nemeobius.
Erycina.
Emesis.
Helicopis.
Euryzona.
Calydna.
Zeonia.

Nymphalidæ.

Pronophila.
Libythea.
Euptychia.
Epinephile.
Erebia.
Chionobas.
Arge = Melanagria.

Mycalesis.
Hipparchia = Satyrus.
Hætera.
Thaumantis.
Brassolis.
Morpho.
Prepona.
Agrias.
Callima.
Apatura.
Adolias.
Charaxes.
Paphia.
Limenitis.
Pyrameis = Cynthia.
Diadema.
Ageronia.
Catagramma.
Vanessa.
Grapta.
Melitæa.
Argynnis.

K

Nymphalis.
Epicalia.
Anartia.
Eunica.
Heterochroa.
Timetes.
Eubagis.
Myscelia.
Acræa.
Colœnis.
Euides.
Heliconia.

Mechanitis.
Ithomia.
Dircenna.
Euplœa.
Danais.

Papilionidæ.

Terias.
Callidryas.
Zegris.
Leptalis.
Colias.

Gonopteryx.
Leucophasia.
Pieris = Pontia.
Anthocharis.
Thyca.
Tachyris.
Callosune.
Iphias.
Thais.
Doritis = Parnassius.
Ornithoptera.
Papilio.

Order VIII. COLEOPTERA.

ELEUTHERATA.

Four wings, the anterior [elytra] hard, meeting down the back by a straight suture; the posterior wings membranous, folded back transversely before the apex. Mandibulate. Four palpi. Larva variable, with legs (six) or apodal; no prolegs. Pupa inactive, showing more or less the parts of the future insect.

The antennæ are generally composed of eleven joints, but sometimes, although very rarely, fifty or more, in *Articerus* only one, varying greatly in character, and occasionally also according to sex. There are two eyes; one or two ocelli are found in some Dermestidæ, as well as in *Homalium* and a few allied genera. That they are true ocelli, however, has been denied. The mouth, very uniform in its type, and complete in its structure, consists of an upper lip or labrum attached to the clypeus, generally by a membrane called the epistome, two strong mandibles, two weaker maxillæ, each carrying a palpus and mostly two-lobed, and a lower lip or labium, with a pair of palpi, and attached to the mentum, which in its turn is attached to the lower part of the head, or jugulum. Some confusion has arisen from calling the labium and mentum together by the former name; the labium then becomes the "ligula," an inappropriate designation, sometimes confounded with "lingula" ["languette" of the French entomologists], and so regarded as a tongue; but if any thing is to be considered in Coleoptera analogous to the tongue, it is the paraglossæ, delicate membranous organs occasionally found behind the labium. The prothorax is the only portion of the thorax seen from above when the elytra are closed, except the scutellum; but this part is frequently wanting; it belongs to the mesothorax.

The elytra, coriaceous, or horny, when there are no inferior wings, are frequently soldered together; in one or two genera one overlaps the other, and in some they diverge. Rarely the females have neither wings nor elytra. The abdomen is connected by its entire anterior portion to the metathorax. A small movable piece attached to the coxa, called the trochantin, is sometimes present. The anterior tarsi are absent in some Lamellicorns, as well as the claw-joint in certain Curculionidæ.

The larvæ vary enormously, from the shrimp-like active *Dyticus* to the footless Curculionidæ. In *Meloë* there are three larval forms.

The cockchafer (*Melolontha vulgaris*), the sacred beetle of the Egyptians (*Scarabæus sacer*), shard or dung-beetle (*Geotrypes stercorarius*), Spanish fly (*Cantharis vesicatoria*), glow-worm (*Lampyris noctiluca*), and corn-weevil (*Calandra granaria*) are members of this order, and, too well known in the larva state, the turnip-fly (*Haltica nemorum*) and others of the same genus, wire-worm (*Agriotes lineatus*), meal-worm (*Tenebrio molitor*), the church-yard beetle (*Blaps mortisaga*), and Colorado potato-beetle (*Doryphora decemlineata*). *Anobium domesticum* has the common name of "death-watch."

Latreille long ago divided this order into four sections, under which the families are even now almost universally arranged. It is, however, a somewhat artificial system, as it would, if strictly followed, widely separate closely allied groups. These sections are dependent on the number of joints of the tarsi, thus:—Pentamera have five joints to all the tarsi; Heteromera have five joints to the four anterior tarsi only, the posterior having only four; Tetramera have the tarsi four-jointed, and Trimera have them three-jointed; but exceptions occur in all. The last two sections have frequently a minute penultimate joint [arthrium], and have therefore been named Subpentamera, Pseudotetramera, and Cryptotetramera, and Subtetramera, Pseudotrimera, and Cyptotrimera respectively. The subsections were ranked as families by Latreille. Stylopidæ, frequently placed in a distinct order [Strepsiptera, or Rhipiptera], are now pretty generally regarded as a degraded type related to Rhipiphoridæ and Meloidæ.

Above 80,000 species belonging to this order are described, distributed under nearly 8000 genera.

TRIMERA.

Maxillary palpi securiform .... APHIDIPHAGA.
Maxillary palpi filiform .... FUNGICOLA.

APHIDIPHAGA.—Maxillary palpi with the last joint securiform. Antennæ short, the last three joints forming a club.

The bodies of these insects are hemispherical ; they have short legs, and strong powers of flight. They feed on Aphides, both in the larval and perfect states.

Coccinellidæ (Lady-birds).	Chilocorus.	Scymnus.
	Coccidula.	Synonycha.
Epilachna.	Cranophorus.	Coccinella.
Exochomus.	Rhizobius.	Megilla.

FUNGICOLA.—Maxillary palpi with the last joint filiform. Antennæ moderately long, flattened, or with a flattened club.

The European species only are known to feed on fungi both in their larval and perfect states. *Trochoideus* has 4-jointed antennæ. Mycetæidæ contains a number of genera of uncertain affinities. Their tarsi are 4-jointed; nevertheless their nearest allies appear to be the Endomychidæ. *Orestia* is referred by Erichson to the Halticidæ. *Trochoideus* is a very aberrant genus, simulating the Paussidæ.

Endomychidæ.	Dapsa.	*Mycetæidæ.*
Stenotarsus.	Corynomalus.	Lithophilus.
Endomychus.	Eumorphus.	Symbiotes.
Lycoperdina.	Encymon.	Leiestes.
Daulis.		Mycetæa.

<table>
<tr><td>Orestia.</td><td>Trochoideus.</td></tr>
</table>

TETRAMERA.

Head rostrate	RHYNCHOPHORA.
Head not rostrate.	
Maxillæ with one lobe	XYLOPHAGA.
Maxillæ with two lobes.	
Antennæ linear.	
Body elongate	LONGICORNIA.
Body ovate or round	PHYTOPHAGA.
Antennæ clubbed at the end	CLAVIPALPI.

CLAVIPALPI.—Last three joints of the antennæ forming a compressed club. Maxillary palpi with the last joint broadly transverse.

These insects are apparently vegetable-feeders ; they are mostly exotic ; the few British species are found in fungi.

Erotylidæ.
Homœotelus.
Zonarius.
Erotylus.
Cyclomorphus.
Tritoma.
Triplax.

Ischyrus.
Mycotretus.
Engis=Dacne.
Thallis.
Triplatoma.
Episcapha.

Encaustis.
Helota.

Languriidæ.

Languria.
Macromela.

PHYTOPHAGA.—Antennæ linear, of moderate length, or short. Body ovate or suborbicular. The elytra covering the sides of the abdomen.

Crioceridæ have an oblong body, and frequently enlarged posterior femora (hence Eupoda of Latreille); many species are more or less aquatic. The other families have mostly a rounded body (Cyclica, Latr.), and are invariably terrestrial. They are all found on plants, feeding principally on the leaves. In this group there are over 10,000 described species.

Cassididæ.
Aspidomorpha.
Elytrogona.
Selenis.
Pœcilaspis.
Chelymorpha.
Mesomphalia.
Coptocycla.
Cassida.
Batonota.
Dolichotoma.
Desmonota.
Tauroma.
Calopepla.
Prioptera.
Himatidium.
Hoplionota.

Hispidæ.
Hispa.
Cephalodonta.
Metaxycera.
Arescus.
Alurnus.
Gonophora.
Cryptonychus.
Hispodonta.

Cephalolia.
Leptispa.
Eurispa.
Aproida.

Galerucidæ.
Cerotoma.
Aplosonyx.
Adimonia.
Galeruca.
Metalepta.
Cœlomera.
Atysa.
Luperus.
Agetocera.
Malacosoma.
Agelastica.
Diabrotica.
Adorium.

Halticidæ.
Psylliodes.
Sphæroderma.
Mniophila.
Linozosta=Hermeo-
    phaga.
Monoplatus.

Octogonotes.
Homotyphus.
Loxoprosopus.
Podontia.
Blepharida.
Œdionychis.
Thyamis=Longitar-
    sus.
Haltica.
Mantura=Balano-
    morpha.
Arsipoda.
Diamphidia.

Chrysomelidæ.
Paropsis.
Phratora.
Timarcha.
Diphyllocera.
Æsernia.
Ceralces.
Doryphora.
Zygogramma.
Chrysomela.
Lina.
Phædon.
Gastrophysa.

Chrysochus.
Corynodes.
Pachnephorus.
Adoxus=Bromius.
Eumolpus.
Colaspis.
Lamprosoma.

Cryptocephalidæ.
Chlamys.
Pachybrachys.

Cadmus.
Monachus.
Cryptocephalus.
Megalostomis.
Euryscopa.
Clytra.
Megalopus.
Mastostethus.
Megascelis.

Crioceridæ.
Crioceris.

Lema.
Hæmonia.
Donacia.
Rhæbus.
Orsodacna.
Ametalla.
Sagra.
Diaphanops.
Carpophagus.
Polyoptilus.
Megamerus.

LONGICORNIA.—Antennæ long, filiform. Body oblong or elon-gate. Female with an ovipositor.

The larvæ are mostly wood-feeders; the perfect insects live but a short time, and feed on vegetable substances; none are known to be aquatic. A few species have the antennæ in the male either pectinate, flabellate, or serrated; the under surface of the joints in many species have one, two, or more pores. Lamiidæ may be almost always recognized by their vertical head; Cerambycidæ have the head porrect; and the Prionidæ have the sides of the prothorax sharply delimited from the upper portion (pronotum), and frequently toothed or spined. *Tmesisternus* and its allies are intermediate forms. *Erichsonia* and *Parandra* differ from all other longicorns, except *Anoploderma,* in having the third tarsal joint entire, not bilobed. There are above 8000 described species in this group.

Lamiidæ.
Amphionycha.
Tetrops=Polyopsia.
Phytœcia.
Glenea.
Saperda.
Agapanthia.
Colobothea.
Exocentrus.
Acanthocinus.
Liopus.
Acanthoderes.
Steirastoma.
Oreodera.
Acrocinus.

Hippopsis.
Oncideres.
Hypselomus.
Compsosoma.
Megabasis.
Pogonochœrus.
Desmiphora.
Symphyletes.
Synelasma.
Praonetha.
Niphona.
Hebesecis = Hebe-cerus.
Zygocera.
Tapeina.

Henicodes.
Olenecamptus.
Petrognatha=Oma-cantha.
Xylorhiza.
Ceroplesis.
Tragocephala.
Sternotomis.
Anoplostetha.
Agelasta.
Gnoma.
Batocera.
Tæniotes.
Monochamus.
Lamia.

Dorcadion.
Parmena.
Microtragus.
Deucalion.
Hexatricha.

———

Tmesisternus.
Pascoea.

Cerambycidæ.

Vesperus.
Distenia.
Coptomma.
Tragocerus.
Trachyderes.
Ctenodes.
Stenygra.
Cosmisoma.
Mydasta.
Collyrodes.
Clytus.
Callidium.
Coremia.
Callichroma.
Deilus.
Pyrestes.
Eroschema.
Distichocera.
Hesthesis.
Tomopterus.
Rhinotragus.
Necydalis.
Cheloderus.

Oxypeltus.
Dorcasomus.
Encyclops.
Leptura.
Ametrocephala.
Macrones.
Anatisis = Petalodes.
Aprosictus = West-
   woodia.
Tricheops.
Obrium.
Ibidion.
Phoracantha.
Eburia.
Phacodes.
Cerambyx.
Cyriopalus.
Metopocœlus.
Torneutes.
Xystrocera.
Œme.
Tetropium.
Spondylis.
Dynamostes.
Thaumasus.

Prionidæ.

Pœcilosoma.
Solenoptera.
Pyrodes.
Anacolus.
Tragosoma.
Prionoplus.
Dœsus.

Philus.
Cyrtonops.
Ægosoma.
Polyoza.
Closterus.
Anacanthus.
Toxeutes.
Hystatus.
Notophysis.
Mallodon.
Rhaphipodus.
Aulacopus.
Macrostoma.
Ergates.
Navosoma.
Callipogon.
Xixuthrus.
Aulacocerus.
Ancistrotus.
Titanus.
Macrodontia.
Acanthophorus.
Prionus.
Dorysthenes.
Prionapterus.
Polyarthron.
Psalidognathus.
Anoploderma.
Sceleocantha.
Cantharocnemis.

———

Erichsonia.

———

Parandra.

XYLOPHAGA.—Head not rostrate. Antennæ short, claviform or perfoliate. Maxillæ with one lobe. Wood-feeders.

The head is sometimes terminated by a short muzzle, but not a true rostrum, the antennæ not being inserted on it, or at any distance from the eyes. In the larval state these insects do immense damage to trees. Bostrychidæ, with which the Xylophaga are sometimes confounded, have, *inter alia*, a two-lobed maxilla.

Scolytidæ.

Tomicus.
Hylesinus.
Platypus.
Scolytus.

RHYNCHOPHORA.—Head prolonged into a rostrum. Antennæ most frequently geniculate, with its basal joint [scape] received into a groove [scrobe]. Vegetable-feeders.

The rostrum varies from a mere vestige to three times the length of the body. The larvæ are without legs; some spin a silky cocoon in which they pass the pupa state. The Curculionidæ are estimated by Jekel to include 30,000 species. It is one of those groups in which many of the forms do not seem to be differentiated into species.

Bruchidæ.

Caryoborus.
Bruchus.
Urodon.

Anthribidæ.

Notioxenus.
Aræocerus.
Brachytarsus.
Cratoparis.
Anthribus.
Eugonus.
Xenocerus.
Apolecta.
Habrissus.
Zygænodes.
Nessiara.
Systaltocerus.
Cedus.
Sintor.
Mecocerus.

——

Xenorchestes.

——

Aglycyderes.

Brenthidæ.

Ulocerus.
Diurus.

Ithystenus.
Brenthus.
Ectocemus.
Arrhenodes.
Ephebocerus.
Taphroderes.
Calodromus.

——

Hypocephalus.

Curculionidæ.

Episus.
Strophosomus.
Naupactus.
Eustales.
Astycus.
Sitona.
Platyomus.
Cyphus.
Præpodes.
Eupholus.
Pachyrhynchus.
Otiorhynchus.
Trachyphlœus.
Phyllobius.
Leptops.
Catasarcus.
Hypsonotus.
Rhigus.
Entimus.

Brachycerus.
Byrsops.
Amycterus.
Psalidura.
Hipporhinus.
Rhyparosomus.
Perperus.
Molytes.
Gonipterus.
Hypera=Phytono-
  mus.
Rhinaria.
Rhadinosomus.
Cleonus.
Lixus.
Hylobius.
Hilipus.
Erirhinus.
Bagous.
Eugnomus.
Tranes.
Oxycorynus.
Belus.
Ctenaphides.
Rhinotia.
Cylas.
Apion.
Apoderus.
Attelabus.
Rhynchites.
Otidocephalus.

Magdalis.
Balaninus.
Anthonomus.
Orchestes.
Tychius.
Sibinia.
Cionus.
Gymnetron.
Læmosaccus.
Haplonyx.
Camarotus.
Alcides.
Cholus.
Homalonotus.
Guioperus.
Acalles.
Cryptorhynchus.
Cratosomus.
Mecomastyx.

Hybomorphus.
Zygops.
Sphadasmus.
Phænomerus.
Arachnopus.
Mecopus.
Talanthia.
Chirozetes.
Odoacis = Macroba-
    mon.
Tachygonus.
Trypetes.
Antliarhinus.
Ulomascus.
Pterocolus.
Ceuthorhynchus.
Eurhinus.
Baris.
Centrinus.

Lyterius.
Rhynchophorus.
Protocerius.
Sphenophorus.
Poteriophorus.
Cercidocerus.
Eugnoristus.
Calandra = Sitophilus
    (Corn-weevil).
Litosomus.
Oxyrhynchus.
Sipalus.
Rhina.
Cossonus.
Glœodema.
Pentarthrum.
Rhyncolus.
Notiomimus.
Onycholips.

HETEROMERA.

Head narrowed behind into a neck......... TRACHELIDA.
Head not narrowed behind .... ATRACHELIA.

TRACHELIDA.—Head exserted, narrowed behind into a neck. Antennæ never clavate (*Tetratoma* excepted). Claws often bifid. In the larval state they feed on fungi, rotten wood, and other vegetable matter; but many are parasitic in the larvæ or in the nests of other insects. In the perfect state they are mostly vegetable-feeders. Meloidæ are wingless, with the edges of the elytra overlapping; their larvæ in the earliest stages are found in the spring on bees and flies, and, being mistaken for lice, were named by Kirby *Pediculus melittæ*. Stylopidæ is the most aberrant family of the Coleoptera; the male only is winged, and ceases to be a parasite only when adult; their victims are almost exclusively bees.

Œdemeridæ.
Mycterus.
Stenostoma.
Œdemera.
Dryops.

Nacerdes.
Agasma.

Cantharidæ.
Goëtymes.
Sitaris.

Cantharis = Lytta
    (Spanish fly).
Palæstra.
Mylabris.
Horia.

Meloidæ.

Cysteodemus.
Meloë.

Stylopidæ.

Stylops.
Xenos.
Hylechthrus.

Rhipiphoridæ.

Rhipidius = Symbius.
Myodites.
Rhipiphorus.
Emenadia.
Evaniocera.
Pelecotoma.

Mordellidæ.

Anaspis.
Mordella.

Pyrochroidæ.

Ischalia = Eupleurida.
Lemodes.
Pyrochroa.

Anthicidæ.

Ochthenomus.
Notoxus.
Formicomus.
Mecynotarsus.

Pedilidæ.

Scraptia.
Macratria.
Pedilus.

Lagriidæ.

Statira.
Emydodes.
Lagria.

Melandryidæ.

Osphya = Nothus.
Stenotrachelus.
Chalcodrya.
Melandrya.
Dircæa.
Phlœotrya.
Serropalpus.
Orchesia.
Mycetoma.
Tetratoma.

Nilionidæ.

Nilio.

Pythidæ.

Agnathus.
Rhinosimus.
Salpingus.
Pytho.

ATRACHELIA.—Head not exserted nor narrowed behind.   Antennæ linear or subclavate.   Claws undivided, in Cistelidæ serrated or pectinated.

The penultimate joint of the tarsi is almost always entire. The typical Tenebrionidæ have mostly connate elytra, and no lower wings. · They are nearly all vegetable-feeders, like the preceding, and invariably terrestrial.   Monommatidæ have been referred to the Erotylidæ.

Cistelidæ.

Prostenus.
Cistela.
Æthyssius = Atractus.

Tenebrionidæ.

Strongylium.
Spheniscus.
Amarygmus.
Megacantha.
Penthe.
Helops.
Cnodalon.

Sphærotus.
Hegemona.
Apocrypha.
Adelium.
Camaria.
Cyphaleus.
Pycnocerus.
Heterotarsus.
Calcar.
Tenebrio.
Eutelus.
Cossyphus.
Cilibe.

Helæus.
Toxicum.
Uloma.
Gnathocerus.
Phrenapates.
Diaperis.
Platydema.
Bolitophagus.
Byrsax.
Phaleria.
Ammobius.
Opatrum.
Praocis.

Sepidium.	Acis.	Mesostena.
Ossiporis.	Morica.	Tentyria.
Moluris.	Stenosis=Tagenia.	Anatolica.
Pimelia.	Steira.	Epiphysa.
Nyctelia.	Eurychora.	Adesmia.
Asida.	Nosoderma.	Erodius.
Embaphion.	Rhypasma.	Arthrodes.
Machla.	Zopherus.	Zophosis.
Elæodes.	Cryptochile.	
Blaps.	Himatismus.	
Scaurus.	Hegeter.	*Monommatidæ.*
Nyctoporis.	Epitragus.	Monomma.

PENTAMERA.

One palpus to each maxilla.
  Elytra covering the abdomen.
    Antennæ pectinated or serrated.
      Prosternum not produced in front ... MALACODERMI.
      Prosternum produced in front .... STERNOXI.
    Antennæ not serrated.
      Palpi prolonged .... PALPICORNIA.
      Palpi shorter.
        Antennæ clavate....................... CLAVICORNIA.
        Antennæ lamelliferous .... LAMELLICORNIA.
    Elytra not covering the abdomen .... BRACHELYTRA.
  Two palpi to each maxilla .... ADEPHAGA.

MALACODERMI.—Prosternum not produced in front, not pointed behind. Antennæ serrated, or produced more or less at the side of the joints. Body frequently soft. Larvæ animal and vegetable feeders; the perfect insects mostly on flowers.

The families of the Malacoderms are variable in their appearance and habits; the name itself is only applicable to the Telephoridæ. The larvæ of the first four families are wood-eaters, but a few feed on animal substances. Some of the Cleridæ, the larvæ of which are mostly animal-feeders, have frequently clavate antennæ; in Rhipiceridæ the antennæ are flabellate or pectinate, with sometimes as many as forty joints. Some of the females of *Lampyris* are larvæform and phosphorescent; they feed on living snails. The phosphorescence, when present, is, in most cases, common to both sexes. *Aspisoma lineatum* is the common fire-fly of the Amazon district; *Luciola italica* of Southern Europe.

Cioidæ.

Lyctus.
Cis.

Bostrychidæ.

Psoa.
Rhizopertha.
Bostrychus = Apate.

Ptinidæ.

Anobium (Death-
    watch).
Xyletinus.
Niptus.
Ptinus.

———

Ectrephes = Ana-
    pestus.

Lymexylidæ.

Atractocerus.
Hylœcetus.
Lymexylon.

———

Cupes.

Cleridæ.

Necrobia.
Corynetes.
Enoplium.
Cormodes.
Stigmatium.
Hydnocera.
Thanasimus.
Trichodes.
Clerus.
Opilus.
Tillus.

Telephoridæ.

Ichthyurus.
Dasytes.
Malachius.
Chalcas.
Melyris.
Drilus.
Malthinus.
Telephorus.
Luciola.
Amydetes.
Dioptoma.

Aspisoma.
Lampyris (Glow-
    worm).
Homalisus.
Lycus.
Eros.
Dictyopterus.

Dascillidæ.

Cyphon.
Scirtes.
Eucinetus.
Eubria.
Dascillus = Atopa.
Lichas.

Rhipiceridæ.

Sandalus.
Rhipicera.
Callirhipis.

———

Cerophytum.

Cebrionidæ.

Cebrio.

STERNOXI.—Prosternum produced in front and pointed behind.
Antennæ filiform or serrated. Vegetable-feeders.

The Elateridæ possess a peculiar process of the prosternum
which, acting as a spring on the mesosternum, enables them, if
placed on their back, to regain their natural position. Their
larvæ are wood- and root-feeders. To *Pyrophorus* belongs some
of the fire-flies of the Tropics. The antennæ of Buprestidæ are
marked by small pores, either diffused, or concentrated in a
depression on each joint.

Elateridæ.

Campylus.
Agriotes.
Corymbites.
Athous.
Dima.
Pyrophorus.

Elater.
Limonius.
Tetralobus.
Semiotus.
Hemirhipis.
Lacon.
Amychus.

Adelocera.

Eucnemidæ.

Microrhagus.
Eucnemis.
Galba.
Anelastes.

Pterotarsus.
Melasis.
Fornax.
Dromæolus.
Tharops.

Throscidæ.
Lissomus.
Drapetes.
Throscus.

Buprestidæ.
Trachys.
Agrilus.
Chrysobothris.
Calodema.
Stigmodera.
Anthaxia.
Curis.
Melobasis.
Buprestis.

Nascio.
Dicerca.
Capnodis.
Polybothris.
Chrysochroa.
Chalcophora.
Euchroma.
Catoxantha.
Julodis.
Sternocera.

LAMELLICORNIA.—Last three joints of the antennæ lamelliform, or in the Lucanidæ pectinate.

The larvæ are wood-eaters, and with the perfect insect also scavengers; many also in the latter state flower- and leaf-eaters. The subfamily Cetoniinæ is often treated as a distinct family; it is differentiated chiefly by the position of the mesothoracic epimera. [The first twelve genera in the following list belong to this subfamily.] Lucanidæ are sometimes ranked as a separate group, under the name of Pectinicornia. In both families there are extraordinary developments of either mandibles, head, or thorax. Trictenotomidæ are an anomalous group, considered by some to be allied to the Prionidæ.

Scarabæidæ.
Valgus.
Trichius.
Gnorimus.
Osmoderma.
Cyclidius.
Cetonia.
Schizorhina.
Pogonotarsus.
Macronota.
Gymnetis.
Heterorhina.
Goliathus.
Syrichthus.
Antedon.
Agaocephala.
Megalosoma.
Chalcosoma.
Dynastes.
Golofa.

Oryctes.
Callicnemis.
Cyclocephala.
Hexodon.
Geniates.
Anoplognathus.
Chrysina.
Chrysophora.
Pelidnota.
Rutela.
Parastasia.
Peperonota.
Antichira = Macraspis.
Anomala.
Phyllopertha.
Anisoplia.
Euchirus.
Pachydema.
Pachypus.

Elaphocera.
Melolontha (Cockchafer).
Ancylonycha.
Rhizotrogus.
Ancistrosoma.
Ceraspis.
Dicrania.
Macrodactylus.
Dichelonycha.
Diphyllocera.
Haplonycha.
Liparetrus.
Calonota=Pyronota.
Phænognatha.
Pachytricha.
Mæchidius.
Diphucephala.
Homaloplia.
Phyllotocus.

Serica.
Hoplia.
Dichelus.
Pachycnema.
Hoploscelis.
Peritrichia.
Glaphyrus.
Amphicoma.
Acanthocerus.
Trox.
Lethrus.
Geotrypes (Dung-
   beetle).
Bolboceras.

Hybosorus.
Aphodius.
Onthophagus.
Bubas.
Phanæus.
Copris.
Canthon.
Scarabæus = Ateu-
   chus.

Lucanidæ.

Sinodendron.
Figulus.
Dorcus.

Leptinopterus = Psa-
   lidostomus.
Lucanus (Stag-
   beetle).
Lamprima.
Dendroblax.
Chiasognathus.
Pholidotus.

Passalidæ.

Passalus.

———

Trictenotomidæ.

Trictenotoma.

CLAVICORNIA.—Antennæ clubbed at the end, the club 2-5-jointed. In the larval and perfect stages living mostly on dead animal matter, several in flowers, or in fungi.

The first three families in the following list are aquatic, except the species of *Sostea.* The majority are known as the " Necrophaga." *Platypsylla* is a remarkable form, a parasite on the Canadian beaver, for which Westwood has proposed a special order—Achrioptera. Le Conte indicates its affinity to several families of this group. Paussidæ are an isolated family; the joints of the antennæ vary from two to ten, all, except the first two when more are present, curiously enlarged.

Heteroceridæ.

Heterocerus.

Parnidæ.

Macronychus.
Elmis.
Parnus.
Sostea.

Georyssidæ.

Georyssus.

Thorictidæ.

Thorictus.

Byrrhidæ.

Chelonarium.
Byrrhus.
Nosodendron.

Dermestidæ.

Anthrenus.
Trogoderma.
Dermestes.
Attagenus.
Byturus.

Cryptophagidæ.

Mycetophagus.
Corticaria.

Monotoma.
Lathridius.
Atomaria.
Cryptophagus.
Antherophagus.

———

Elacatis = Othnius.

Cucujidæ.

Silvanus.
Brontes.
Inopeplus = Ino.
Hemipeplus =
   Ochrosanis.
Palæstes.

Cucujus.
Ancistria.
Hectarthrum.
Passandra.

Colydiidæ.

Cerylon.
Dastarcus.
Pycnomerus.
Bothrideres.
Deretaphrus.
Aglenus.
Nematidium.
Colydium.
Acropis.
Aulonium.
Synchita.
Bitoma.
Cossyphodes.
Tarphius.
Enarsus.
Endophlœus.
Sarrotrium.

Rhysodidæ.

Rhysodes.
———
Omma.

Trogositidæ.

Thymalus.
Peltis.
Elestora.
Leperina.
Gymnochila.
Temnochila.
Trogosita.
Melambia.
Nemosoma.

Nitidulidæ.

Rhizophagus.
Ips.
Paromia.
Cychramus.
Camptodes.
Lordites.
Meligethes.
Omosita.
Nitidula.
Brachypeplus.
Carpophilus.
Conotelus.
Mystrops.
Cercus.

Phalacridæ.

Olibrus.
Tolyphus.
Phalacrus.

Histeridæ.

Acritus.
Abræus.
Onthophilus.
Plegaderus.
Trypanæus.
Saprinus.
Carcinops.
Hetærius.
Margarinotus.
Hister.
Platysoma.
Hololepta.
Murmidius.

Scaphidiidæ.

Scaphisoma.
———
Paussidæ.

Paussus.
Hylotorus.

Scaphidium.
Diatelium.

Trichopterygidæ.

Ptilium.
Trichopteryx.
Limulodes.

Platypsyllidæ.

Platypsylla.

Corylophidæ.

Alexia.
Sacium.
Corylophus.

Anisotomidæ.

Clambus.
Agathidium.
Anisotoma.
———
Sphærius.

Silphidæ.

Catops = Choleva.
Leptinus.
Adelops.
Silpha.
Necrophorus.
Leptodirus.

Scydmænidæ.

Mastigus.
Clidicus.
Scydmænus.

Gnostidæ.

Gnostus.

BRACHELYTRA.—Elytra very short, not covering the abdomen. Antennæ short, never clubbed. Two anal appendages. Voracious animal-feeders, occasionally living on decaying vegetable matter or carrion. Larva resembling the imago.

In Staphylinidæ the abdomen is free and at the tip furnished with vesicular papillæ connected with glands secreting a disagreeable fluid. Some species reside in ants' nests; *Velleius dilatatus* in the nests of hornets. Pselaphidæ are trimerous, have the abdomen fixed, and without anal appendages. They live in moss, and many in ants' nests. In *Articerus* each antenna consists of a single joint.

Staphylinidæ.

Micropeplus.
Anthobium.
Homalium.
Lesteva.
Anthophagus.
Micralymma.
Piestus.
Syntomium.
Oxytelus.
Bledius.
Osorius.
Platystethus.
Stenus.
Pæderus.
Sunius.
Lithocharis.
Stilicus.

Lathrobium.
Oxyporus.
Astrapæus.
Quedius.
Velleius.
Philonthus.
Ocypus.
Staphylinus.
Xantholinus.
Othius.
Mycetoporus.
Tachinus.
Conosoma = Conurus.
Hypocyptus.
Diglossa.
Dinarda.
Lomechusa.
Spirachtha.

Atemeles.
Aleochara.
Oxypoda.
Homalota.
Myrmedonia.
Autalia.

Pselaphidæ.

Claviger.
Articerus.
Euplectus.
Bryaxis.
Amaurops.
Pselaphus.
Chennium.
Ctenistes.
Tyrus.
Faronus.

PALPICORNIA.—Maxillary palpi elongate. Antennæ short, generally club-shaped. Mostly aquatic; in the perfect state herbivorous.

In some of the Hydrophilidæ the palpi are longer than the antennæ. Sphæridiidæ are mostly found in excrementitious matter.

Sphæridiidæ.

Cercyon.
Sphæridium.

Hydrophilidæ.

Ochthebius.

Hydrochus.
Hydræna.
Helophorus.
Spercheus.
Globaria.
Berosus.
Limnebius.

Laccobius.
Hydrobius.
Rygmodus.
Hydrous.
Hydrophilus.
Philhydrus.
Tropisternus.

ADEPHAGA.—Two palpi to each maxilla. Antennæ filiform. In the larval and perfect states voracious animal-feeders.

The first two families are exclusively aquatic (Hydradephaga), but flying occasionally at dusk. Carabidæ and Cicindelidæ, except a few of both families, which live partially under water, but never swimmers, are terrestrial (Geodephaga). Gyrinidæ have four eyes.

Gyrinidæ.
Porrorhynchus.
Gyrinus (Whirlgig).

Dyticidæ.
Dyticus=Dytiscus.
Colymbetes.
Hydroporus.
Pelobius.
Haliplus.
————
Amphizoa.

Carabidæ.
Bembidium.
Cillenum.
Ega.
Aëpus.
Anophthalmus.
Pogonus.
Dyscolus.
Anchomenus.
Calathus.
Sphodrus.
Metius.
Amara.
Zabrus.
Feronia.
Catadromus.
Trigonotoma.
Stenolophus.
Harpalus.
Barysomus.
Acinopus.
Diachromus.
Cratocerus.
Daptus.

Cyclosomus.
Pelecinus.
Promecognathus.
Stomis.
Broscus.
Miscodera.
Dioctes.
Badister.
Licinus.
Oodes.
Chlænius.
Callistus.
Loricera.
Panagæus.
Clivina.
Scarites.
Carenum.
Pasimachus.
Morio.
Hyperion=Campylocnemis.
Anthia.
Graphipterus.
Apotomus.
Ditomus.
Siagona.
Ozæna.
Adelotopus.
Silphomorpha.
Mormolyce.
Scopodes.
Catascopus.
Dromius.
Cymindis.
Agra.
Brachinus.
Helluo.

Zuphium.
Galerita.
Drypta.
Casnonia.
Hexagonia.
Pamborus.
Tefflus.
Cychrus.
Scaphinotus.
Damaster.
Carabus.
Haplothorax.
Calosoma.
Leistus.
Nebria.
Elaphrus.
Notiophilus.
Omophron.

Cicindelidæ.
Ctenostoma.
Procephalus.
Pogonostoma.
Collyris.
Tricondyla.
Therates.
Euprosopus.
Dromica.
Odontochila.
Cicindela.
Eurymorpha.
Oxygonia.
Megacephala.
Oxychila.
Omus.
Platychile.
Manticora.

L

Order IX. HYMENOPTERA.

Four naked, membranous, unequal few-veined wings. Mandibulate; labium and maxillæ more or less elongate. Larvæ apodal [excl. Terebrantia]. Pupa incomplete, inactive.

The mouth has two robust mandibles, protected by a labrum above, and two united maxillæ forming a sheath for the labium, when developed into a suctorial proboscis, below, and two pairs of palpi. The mandibles, however, are chiefly employed in the construction of the nest; the labium, when used as a sucking-organ, is retractile.

The abdomen, mostly attached by a pedicel to the thorax, is in the female furnished with an ovipositor, which is frequently modified into a sting [aculeus], or in one group into saws [serræ], and in another into a borer [terebra]. It consists normally of six pieces, the two outer forming a sheath. The mesothorax forms the bulk of the thorax, the other two parts being but moderately developed. The wings are always horizontal, the upper pair frequently with a chitinous nodule (tegulum) at the base; the lower pair along its anterior margin is provided with a series of minute hooks, which catch into the reflected posterior margins of the upper wings, for the purpose of keeping them together during flight. On the upper wings is an opaque spot, as in the Neuroptera, named the "stigma."

Wingless species (but the absence of wings frequently depending on sex) are frequent. Female ants, after pairing or when preparing to lay their eggs, *shake off* their wings, a small portion, however, generally remaining. The wings, as in Chalcididæ, are sometimes almost destitute of venation. Some species are eyeless, or are without mandibles, or without a sting. Nevertheless the order cannot be said to contain any aberrant or abnormal forms, except among the neuters; these are undeveloped females peculiar to the Hymenoptera, with the sole exception of white ants among the Neuroptera.

In the aculeate Hymenoptera a poison, strongly impregnated with formic acid, is secreted by a special gland.

Galls, although produced by other insects, and even by worms, are in this order almost exclusively caused by Cynipidæ. These insects deposit their eggs in the leaves, stems, and roots of plants. Different kinds of galls are seen on the same plant, produced by different species of gall-flies. Some galls contain but a single egg, as those of *Cynips gallæ-tinctoriæ*; in other galls more than a thousand eggs will be found, as in *Cynips quercus-radicis*. Oak-apples are the galls of *C. terminalis*; the bullet-galls of the oak, of *Cynips Kollari*; and oak-spangles, of *C. longicornis*. The "bedeguar," or gall of the wild rose, is caused by *Rhodites rosæ*. *Cynips aptera* form clusters of galls on the roots of the oak. Gall-like substances are also produced by other Hymenoptera.

In the Terebrantia the larvæ resemble the caterpillars of the Lepidoptera, having legs varying from three to eleven pairs, but always six true legs, the remainder abdominal or prolegs. All other Hymenoptera have footless larvæ, which are often parasitic. Many species of Hymenoptera spin cocoons.

To this order belong saw-flies (*Tenthredinidæ*), the turnip saw-fly (*Athalia spinarum*), whose black-caterpillar is so destructive to our turnip-crops, ants (*Formicidæ*), the wasp (*Vespa vulgaris*), hornet (*Vespa crabro*), the honey-bee (*Apis mellifica*), and humble-bees (*Bombi*). More than 16,000 species are known, of which about 3000 are British.

Six groups may be distinguished; the last two or three are frequently mentioned as the "aculeate Hymenoptera.

Posterior segments of the abdomen retractile ...... TUBULIFERA.
Posterior segments of the abdomen not retractile.
   Females armed with a saw or borer.
      Abdomen sessile .... TEREBRANTIA.
      Abdomen petiolate .... PUPIVORA.
   Females armed with a sting.
      With wingless neuters........................... HETEROGYNA.
      With winged neuters, if any.
         Basal joint of the posterior tarsi cylindrical FOSSÓRES.
         Basal joint of the posterior tarsi dilated ... MELLIFERA.

TEREBRANTIA.—Abdomen sessile, in the female armed with a saw or saws or a borer. Larva with six legs and with a number of prolegs. Feed on vegetable substances.

By means of the saw or borer the female makes slits or punctures in the shoots of young plants or in trees, in which she places her eggs; galls are frequently the result. A number of species form double cocoons, sometimes of earth, in which they pass the winter. In the Tenthredinidæ the female is furnished with a pair of saws, in the Siricidæ with a borer.

Tenthredinidæ.	Perga.	Fenusa.
Cimbex.	Hylotoma.	Selandria.
Abia.	Pachylota.	Allantus.
Trichiosoma.	Athalia.	Emphytus.

Tenthredo (Saw-	Crœsus.	*Siricidæ.*
fly).	Lophyrus.	Cephus.
Cladius.	Tarpa.	Sirex=Urocerus.
Nematus.	Lyda.	Xiphydria.
Dolerus.	Xyela.	Oryssus.

PUPIVORA.—Abdomen petiolated, in the female armed with a saw or borer. Larva without feet.

The females deposit their eggs in the larvæ or pupæ of other insects, or, in the Cynipidæ, in plant-structures or in galls. Cynipidæ are mostly gall-makers, but some, *e. g. Figites, Allotria,* &c., are insect-parasites like the other families of the group. Braconidæ (=Ichneumones adsciti) are differentiated from the genuine Ichneumones by having only one recurrent nerve in the fore wing, instead of two. The Pupivora exactly correspond to the Spiculifera of Westwood.

Cynipidæ (Gall-flies).
Ibalia.
Ægilips.
Eucœla.
Figites.
Cynips.
Teras.
Biorhiza.
Rhodites.
Andricus.
Synergus.

Chalcididæ.
Chalcis.
Halticella.
Leucospis.
Agaon.
Apocrypta.
Sycophaga.
Torymus.
Monodontomerus.
Callimome.
Eurytoma.
Eucharis.
Perilampus.
Ormyrus.
Encyrtus.

Myina.
Eupelmus.
Trigonoderus.
Chiropachus.
Cleonymus.
Pteromalus.
Eunotus.
Melittobia.
Entedon.
Cirrospilus.
Trichogramma.
Omphale.
Entedon.
Eulophus.

Proctotrypidæ.
Helorus.
Belyta.
Cinetus.
Paramesius.
Proctotrypes.
Embolimus.
Gonatopus.
Chelogynus.
Antæon.
Bethylus.
Megaspilus.

Ceraphron.
Telenomus.
Teleas.
Prosacantha.
Platygaster.
Mymar.
Cosmocoma.
Prestwichia.

Braconidæ.
Dacnusa.
Alysia.
Opius.
Perilitus.
Aphidius.
Microgaster.
Chelonus.
Sigalphus.
Rhogas.
Colastes.
Bracon.

Ichneumonidæ.
Lissonota.
Glypta.
Pimpla.
Ephialtes.

Bassus.
Tryphon.
Banchus.
Exetastes.
Porizon.
Campoplex.
Paniscus.

Ophion.
Agriotypus.
Pezomachus.
Hemiteles.
Cryptus.
Amblytelus.
Ichneumon.

Evaniidæ.

Stephanus.
Evania.
Pelecinus.
Fœnus.
Megalyra.
Trigonalys.

TUBULIFERA.—Posterior segments of the abdomen retractile, and provided with a membranous ovipositor, composed of a single piece. Solitary, deposit their eggs in the nests of other Hymenoptera.

Richly coloured insects, only appearing in the hottest sunshine, and capable of rolling themselves up into a ball. They are remarkable for having the underside of the abdomen concave.

Chrysididæ.

Chrysis.
Euchrœus.
Stilbum.
Homalus.

Elampus.
Hedychrum.
Parnopes.
Cleptes.

HETEROGYNA.—Males, females, and neuters; the latter abortive wingless females, sometimes of two kinds—workers and soldiers. Social, build nests, and excavate tunnels.

The soldiers (or "policemen") have very large heads. The females and workers of many species are furnished with a sting. Some species carry off the larvæ of other species, which are educated as slaves and then take charge of their captor's colony. The common *Formica sanguinea* is one of the slave-making ants.

The Dorylidæ are differentiated from the Formicidæ by the first segment of the abdomen only forming the peduncle. Their sexes are not certainly known.

Formicidæ (Ants).

Formica.
Tapinoma.
Ponera.
Polyergus.
Odontomachus.
Polyrhachis.
Ectatomma.
Œcophylla.
Myrmica.

Myrmecia.
Myrmecina.
Stenomma.
Crematogaster.
Eciton.
Solenopsis.
Strongylognathus.
Daceton.
Phidole.
Monomorium.
Atta.

Œcodoma.
Cryptocerus.

Dorylidæ.

Anomma.
Myrmecocystus.
Labidus.
Dorylus ♂
Dichthadia ♀ } ?
Typhlopone, ☿

FOSSORES.—Neuters, if any, winged. Basal joint of the posterior tarsi cylindrical. A sting.

The neuters are mostly, perhaps entirely, confined to the Vespidæ. Nearly all the species are fossorial; but there is no special adaptation of legs, except that in some the anterior tibiæ and tarsi are ciliated. They burrow in rotten wood, twigs, sand banks, or construct mud or paper nests. The larvæ often feed on other larvæ, flies, spiders, &c., provided by the mother.

The females of Mutillidæ are wingless. The upper wings in Vespidæ (= Diploptera) are folded longitudinally.

Mutillidæ.
Mutilla.
Myrmosa.
Methoca.
Tiphia.
Thynnus, ♂
Myrmecoda, ♀ }

Scoliidæ.
Myzine.
Elis.
Scolia.
Sapyga.

Bembecidæ.
Bembex.
Stictia = Monedula.

Vespidæ.
Vespa (Wasp).
Polistes.
Polybia.
Eumenes.
Odynerus.
Synagris.
Epipone.
Abispa.
Rhynchium.
Celonites.
Masaris.
Paragia.
Apoïca.

Crabronidæ.
Crabro.
Oxybelus.
Trypoxylon.
Psen.
Pemphredon.
Dinetus.
Miscophus.
Passalœcus.
Stigmus.
Nysson.
Larra.
Gorytes.
Pison.
Tachytes.
Astata.
Mellinus.
Cerceris.
Philanthus.

Sphegidæ.
Sphex.
Ampulex.
Trigonopsis.
Ammophila (Sand-wasp).
Miscus.
Pelopœus.
Chlorion.

Pompilidæ.
Pompilus.
Dolichurus.
Aporus.
Pepsis.
Mygnimia.
Ceropales.

MELLIFERA (Bees).—Neuters winged. Basal joint of the posterior tarsi dilated, adapted for collecting and carrying pollen. A sting in the females and neuters.

The only genera having neuters or workers in this country are *Apis* and *Bombus*. The basal joint of the posterior tarsi is scarcely dilated in the solitary species. These deposit their eggs either in the nests of other species whose young are starved by the intruders, or they form nests in twigs, rotten wood, old walls, banks, &c. Social species construct combs, composed of hexagonal cells, for storing the food [honey] of their larvæ. Honey is the nectar collected from flowers which, undergoing a certain alteration in the stomach, is regurgitated into the cells. Wax is digested honey emitted through the sides of the ventral segments and worked up by the mandibles into a state fit to form the cells. The queen-bee is a selected and richly-fed larva; during her life of five years she is said to lay about a million of eggs. Tegetmeier cites cases of eggs laid by neuters. The unimpregnated eggs give rise to drones, the impregnated eggs to females (and neuters). What is called the wedding-flight, which occurs but once in a lifetime, is always taken in the air.

A community of the hive bee is estimated by Westwood to contain 2000 males or drones, 50,000 workers, and 1 queen.

Pollen is collected either with the hind femora (*Apis, Andrena*), with the abdomen (*Megachile, Osmia, Anthidium*), or with the hind tibiæ (*Anthophora, Eucera*). *Sphecodes* and *Prosopis* do not collect. *Melecta, Nomada, Epeolus*, and others are solitary and parasitical. Andrenidæ have short tongues, Apidæ long ones. There are about 250 British species.

Apidæ.		*Andrenidæ.*
Apis (Hive-bee).	Oxæa.	Stelis.
Melipona.	Lestis.	Cœlioxys.
Trigona.	Xylocopa.	
Bombus (Humble-bee).	Megachile.	*Andrenidæ.*
Eulema.	Chalicodoma.	Panurgus.
Apathus = Psithyrus.	Osmia.	Andrena = Melitta.
Euglossa.	Ceratina.	Cilissa.
Aglaë.	Heriades.	Dasypoda.
Epicharis.	Chelostoma.	Nomia.
Centris.	Anthidium.	Cyathocera.
Anthophora.	Melecta.	Halictus = Hylæus.
Saropoda.	Chrysantheda.	Colletes.
Eucera.	Thalestria.	Macropis.
Systropha.	Crocisa.	Augochlora.
	Nomada.	Agapostemon.
	Epeolus.	Sphecodes = Dichroa.
		Prosopis.

Subkingdom VI. **MOLLUSCA.**

HETEROGANGLIATA.

Soft-bodied, unsegmented animals, often protected by a shell. The digestive system including stomach, intestine, and anus. Reproduction by ova, rarely by gemmation. Hermaphrodite, but generally the sexes on distinct individuals [diœcious]. A metamorphosis in nearly all.

The nervous system generally consists of three pairs of ganglia, giving off branches to the different parts; but in Brachiopoda and Tunicata there is only one ganglion. The heart has two or more chambers, but in the Ascidians it is reduced to a simple tube. The blood is generally colourless, and circulates in sinuses, having no proper walls. Respiration is mostly effected by branchiæ, a specialized portion of the mantle; there is one or, most commonly, two on each side.

The integument of the body, continuous or divided into two lobes, is called the mantle [pallium]. From the ventral surface projects a muscular disk or foot [podium], a modification of the lower lip and an organ of locomotion, and generally the only one, but it is sometimes absent. The foot is often divisible into three parts—pro-, meso-, and metapodium; it is the last that sometimes secretes a calcareous or horny disk, known as the " operculum."

The mantle secretes the shell, which is rarely absent; it may be in one or two pieces [univalve or bivalve], rarely of many pieces [multivalve]; in a few instances it is internal. The inner layer of the shell is sometimes nacreous in its texture; its colour is due to finely sculptured lines. A rudimentary shell is frequently present whilst the young mollusk is still in the egg, and it has even been observed in forms which have no shell in the adult state.

In certain classes the mouth is provided with an organ [radula or odontophore], often, but erroneously, called the tongue. It is armed with teeth (in the large garden-slug, *Limax maximus*, they amount to 26000) and is an important part in classification. It is found in Pteropoda, Gastropoda, and Cephalopoda.

The embryo of the Mollusca ordinarily passes through three stages :—(1) the *Gastrula*-stage, including the earlier *Morula*; (2) the *Trochosphæra*-stage, when the embryo is girdled with a row of cilia; and (3) the *Veliger*-stage, when it has acquired the "velum"—a ciliary cephalic expansion of the integument, said to be identical with the ciliated disk of the Rotifera. The ova are often defended by hard albuminous capsules of various, and sometimes very complex, forms.

"In the Mollusca the developmental energies seem to have been expended chiefly in the perfection of the vegetative series of organs, or those concerned in the immediate preservation of the individual and the species."

The limits of the Mollusca are at present unsettled. Ray Lankester places the Tunicata with the Vertebrata, Häckel and other biologists with the Vermes. Molluscoida included Polyzoa, Brachiopoda, and Tunicata. A "roughly" drawn distinction is sometimes made between those Mollusca with a head [Encephala or Cephalophora] and those without [Acephala]. In the former the head is generally provided with tentacula, eyes, and a mouth armed with jaws; in the latter there is no cephalic ganglion, and the mouth is a simple inlet for the food, "having no power of selection in the first instance." Otocardia, excluding Brachiopoda and Tunicata, is another name for the Mollusca. Schmarda has seven classes, including Polyzoa; Claus four, excluding Brachiopoda and Tunicata,—Gastropoda including Heteropoda and Pteropoda, while Scaphopoda is raised to the rank of a class.

Animal enclosed in a bivalve shell.
    Mouth with two arms...................... BRACHIOPODA.
    Mouth without arms .... LAMELLIBRANCHIATA.
Animal either naked or enclosed in a univalve shell.
    With a head; heart with two or more chambers.
        Mouth with long arms .... CEPHALOPODA.
        Mouth without arms.
            Locomotion effected by a ventral disk .... GASTROPODA.
            Locomotion effected by fin- or wing-like appendages.
                With a fin-like tail or a ventral fin .... HETEROPODA.
                With two wing-like expansions... PTEROPODA.
    Without a head; heart a simple tube ... TUNICATA.

Class I. BRACHIOPODA.

SPIROBRANCHIATA. PALLIOBRANCHIATA. HAPLOCARDIA.

Headless, symmetrical mollusks, enclosed in a bivalve shell. Mouth with two long, cirriferous arms. No true branchiæ. Mostly hermaphrodite.

The valves of the shell are above and below the animal, not right and left, as in ordinary bivalves, and they have no elastic ligament; the lower, posterior, or ventral valve is frequently prolonged into a beak, and perforated, to allow the passage of a peduncle by which the animal attaches itself to some foreign body. The dorsal integument lines the interior of the valves, and "forms by its expansions the lobes of the mantle," here subservient to respiration. The heart is a simple ventricle; the cavity formed by the lobes is partly occupied by "two long fringed arms, continued from the sides of the mouth, and disposed in folds and spiral curves."

Most of the species (upwards of 1800) are extinct; they appear to have been most abundant in the Silurian epoch; about 80 still exist, some living at a depth of 2500 fathoms.

Brachiopoda are embryologically allied to the Vermes, and by some writers they are cousidered to be very similar in structure to the Polyzoa. Ray Lankester unites them with the Lamellibranchiata, under the name of Lipocephala. Claus places them as a pendant to the Mollusca. They are the lowest "stage of genuine mollusks," according to Häckel; but others consider that they are most allied to Annelida. Huxley holds their affinities to be between the Polyzoa and the higher Mollusca.

There are two orders :—

Shell hingeless .... LYOPOMATA.
Shell with a hinge-line .... ARTHROPOMATA.

Order I. LYOPOMATA.

ECARDINES. PLEUROPYGIA. INARTICULATA. SARCOBRANCHIATA.

Valves not toothed, held together by muscles; shell corneous. Intestine terminating in a lateral anus.

The oral arms are mostly fleshy, and without the bony support of the next order. In the Craniidæ the shell is calcareous; in the Lingulidæ it is covered with a horny epidermis; in the latter the peduncle is emitted from between the valves.

Of the three living genera only a very few species are known; one—*Crania*, with two species—is found in European seas. *Lingula anatina* and two or three other species, now confined to the Eastern seas, do not differ morphologically from their congeners of the Cambrian epoch. *Discina*, of which there are several species, is found in China, the West Indies, and the Pacific coasts; it was also existent in the Silurian epoch.

Lingulidæ.	*Discinidæ.*	*Craniidæ.*
Lingula.	Discina.	Crania.
*Obolus.	*Trematis.	

Order II. ARTHROPOMATA.

TESTICARDINES. APYGIA. ARTICULATA. SCLEROBRANCHIATA.

Valves held together by teeth. Shell calcareous. Intestine without an anus.

A peculiar calcareous loop is attached to the upper valve of most of the species, destined for the support of the arms. The alimentary canal is short, with a cæcal end.

The recent species, mostly confined to deep water, are found in almost all parts of the world. The Terebratulidæ are commonly known as "lamp-shells."

Terebratulidæ.	*Spiriferidæ.*	*Orthidæ.*
Terebratula.	Spirifer.	*Orthis.
Waldheimia.	*Koninckia.	
Kraussia.	*Spirigera.	*Productidæ.*
Argiope.		*Productus.
Thecidium.	*Rhynchonellidæ.*	*Chonetes.
	Rhynchonella.	

Class II. LAMELLIBRANCHIATA.

BIVALVIA. CONCHIFERA. ACEPHALA. DITHYRA. CORMOPODA.

Headless mollusks encased in a bivalve shell, sometimes with accessory valves. Body enclosed within a mantle. Respiratory organs consisting of lamelliform or filamentous branchiæ. Sexes distinct.

The branchiæ are usually two on each side, placed between the mantle and the foot, and well supplied with cilia. The mouth has no radula. The heart is always well developed. Ocelli are present in most.

The shell is frequently inequivalve and inequilateral, with either one or two adductor muscles for closing its valves. The apex of each valve is the umbo, and between them is the hinge [cardo], with or without teeth, where the two valves are joined to one another.

Locomotion is very imperfect in the adult state ; many are per-manently fixed, either by their shells or by a peculiar secretion, one form of which is known as a "byssus," or they bury them-selves in the sand, or bore into timber or rocks. They are mostly very prolific ; in the oyster the number of ova [known as the "spat"] varies from 250,000 to 800,000 in the season ; but they take three years to come to maturity. The young of this class are in their earliest stage ciliated and free-swimming.

While Huxley sees nothing having the value of orders in this class, Carus gives us ten. Schmarda has five orders divided into two sections—Endocardines (Rudistæ only) and Exocardines (Monomya, including *Ostrea, Pecten*, &c. ; Heteromya, *Mytilus*, &c. ; Isomya, *Arca, Unio, Chama, Venus*, &c. ; and Inclusa, *Gastrochæna, Pholas*, &c.). Some writers adopt two divi-sions, depending on the number of adductor muscles, those bi-valves with one adductor being called "Monomyaria," and those with two "Dimyaria ;" or, again, we have two divisions, based on a character nearer ordinal rank than the other—viz. the absence or presence of siphons (tubular prolongations of the mantle).

There is said to be 14,000 species, recent and fossil.

With siphons.. SIPHONIATA.
Without .... ASIPHONIATA.

Order I. ASIPHONIATA.

ATRACHIA.

Headless mollusks without respiratory siphons. Lobes of the mantle free.

Most of the Asiphoniata are fixed and motionless, the foot being either small or wanting ; in the former case frequently provided with a byssus-secreting gland, situated at the base of the foot, by which it attaches itself to foreign bodies.

This order includes the oyster (*Ostrea edulis*), pearl-oyster [of Indian seas] (*Meleagrina margaritifera*), pearl-oyster [of European rivers] (*Margaritana margaritifera*), mussel (*Mytilus edulis*), and scallop (*Pecten maximus*). Freshwater mussels are various species of *Unio* and *Anodonta*.

Anomiidæ.	*Limidæ.*	*Arcidæ.*
Anomia.	Lima.	Leda.
Placunanomia.		Nucula.
	Aviculidæ.	Axinæa = Pectun-
Placunidæ.	Malleus.	culus.
Placuna = Placenta.	Vulsella.	Oucullæa.
	Perna.	Anomalocardia.
Ostreidæ.	Meleagrina.	Arca.
Gryphæa.	Avicula.	
Ostrea (Oyster).		*Trigoniidæ.*
	Pinnidæ.	Trigonia.
Spondylidæ.	Pinna.	
Spondylus.		*Unionidæ.*
	Mytilidæ.	Ætheria.
Pectinidæ.	Prasina.	Hyria.
	Dreissena.	Mycetopus.
Amusium = Pleuro-	Mytilus (Mussel).	Anodonta.
nectia.	Modiola.	Margaritana.
Pecten (Scallop).	Lithodomus = Litho-	Unio.
	phagus..	

Order II. SIPHONIATA.

MACROTRACHIA.

Headless mollusks with one or two respiratory siphons. Lobes of the mantle more or less united.

Locomotion, effected by the more developed foot, is general in this order, and the respiratory system is more complex. For this purpose there are one or two tubular prolongations [siphons] from the margin of the mantle ; when two, generally united—the one inhalant, the other exhalant, and furnished with muscular fibres for their emission and retraction, but in some cases the siphon is incapable of being retracted. The foot is sometimes provided with an orifice through which the water enters, and when swollen by its admission the foot may exceed the capacity of the shell.

This order includes the clam [of North America] (*Mactra solidissima*) [the river-clams are *Unios*], cockle (*Cardium cdule*), cob (*Mya arenaria*), razor-shell (*Solon siliqua*), and ship-worm (*Teredo navalis*). The shell of *Tridacna gigas*, one valve of which serves as a font in many continental churches, frequently attains a weight of 500 pounds.

Pholadidæ have accessory valves in the hinge of the shell. In *Teredo* the valves are merely appendages of the foot. In Gastrochænidæ the valve or valves are adherent to a tubular shell.

The extinct family Hippuritidæ (= Rudistæ) is characterized by an inner 1–3-toothed hinge, the teeth confined to one valve. It is a distinct order for some authors; allied to Chamidæ for others. It is only known in the Cretaceous period.

Chamidæ.
Chama.
Tridacna.
Hippopus.

Cardiidæ.
Didacna.
Cardium (Cockle).

Lucinidæ.
Solenomya.
Kellia.
Cyanium = Turtonia.
Cryptodon.
Corbis.
Lucina = Loripes.

Cyrenidæ.
Pisidium.
Cyclas = Sphærium.
Cyrena.

Isocardiidæ.
Cyprina.
Isocardia.

Astartidæ.
Cardita.
Mytilicardia.
Astarte.
Gouldia.
Galeomma.
Scintilla.

Veneridæ.
Artemis.
Meroë.
Cytherea.
Venus.
Tapes.
Petricola.
Venerupis.

Mactridæ.
Lutraria.
Mactra.

Tellinidæ.
Donax.
Asaphis.
Tellina.
Psammobia.

Solenidæ.
Solecurtus.
Solen (Razor-shell).

Myidæ.
Mya (Cob).
Corbula.
Neæra.
Panopæa.

———

Hippuritidæ.
*Hippurites.
*Rudistes.
*Caprina.

Anatinidæ.
Glauconome.
Pholadomya.
Anatina.
Lyonsia.

Chamostreidæ.
Chamostrea = Cleidothærus.

Amphidesmidæ.
Amphidesma.
Ervilia.
Paphia.
Thracia.
Pandora.

Gastrochænidæ.
Gastrochæna.
Brechites = Aspergillum.

Saxicavidæ.
Saxicava.
Glycimeris.

Pholadidæ.
Pholadidia.
Pholas (Piddock).

Teredinidæ.
Teredo (Ship-worm).

Class III. PTEROPODA.

COPONAUTÆ.

Free pelagic mollusks swimming by means of two fin-like expansions developed from the anterior extremity. Hermaphrodite.

The head, not well defined, is expanded on each side into a large muscular fin [epipodium]. There is a small mouth, sometimes tentaculate and having a radula. The mantle may be absent or only slightly developed, or it is only present in the earlier stages. There is no proper respiratory organ, but there is occasionally a ciliated branchial sac.

These are small, active animals, gaily coloured, mostly provided with thin, symmetrical shells, and found in large numbers on the surface at night (Major Owen), or glistening in the sunshine (Wyville Thomson). They "absolutely swarm" in the high seas; in the north *Clio borealis* and *Limacina arctica* are the chief food of the whale.

Pteropoda are Schmarda's twenty-sixth class. For Claus they are a subclass of Gastropoda. According to the former there are scarcely 100 living, and about 150 fossil species.

Without a shell .... GYMNOSOMATA.
With a shell .... THECOSOMATA.

Order I. GYMNOSOMATA.

Head and foot distinct; no mantle. Shell absent. Larva eventually with cilia.

The fins are attached to the neck; the head is tentaculate, except in *Cliodita*, and in *Pneumodermon* the mouth also. The latter has a rudimental shell placed at the bottom of the visceral cavity.

Cliidæ.	Cliodita.	*Pneumodermidæ.*
Clio.	Clionopsis. Eurybia.	Pneumodermon.

Order II. THECOSOMATA.

Head rudimentary. Shell always present, but internal in Cymbuliidæ. Larva without cilia.

In these animals the hinder part of the body is protected by a light transparent or semitransparent shell, variously shaped, but spiral in Limacinidæ.

Of the fossil species, many are congeneric with existing forms, and some of them are of comparatively large size, *e.g. Conularia*, which is sometimes nearly two feet in length. The Silurian *Tentaculites*, "though often referred to the Tubicolar Annelides, appear to belong, without doubt, to the Pteropoda."

Hyalæidæ.	Hyalæa=Cavolina.	*Thecidæ.*
Triptera.	*Cymbuliidæ.*	*Conularia.
Cleodora.		
Balantium.	Cymbulia.	*Limacinidæ.*
Styliola=Creseis.	Halopsyche=Psyche.	Limacina.
Diacria.	Tiedemannia.	Spirialis.
	*Tentaculites.	

Class IV. GASTROPODA.

BRANCHIOGASTEROPODA.   COCHLIDES.   PLATYPODA.

Land or water mollusks, generally encased in a univalve shell. Locomotion effected by the ventral disk or foot. A distinct head in nearly all, with one or two pairs of tentacles. Diœcious or hermaphrodite.

A heart, liver, and convoluted intestine are mostly present: the mouth is provided with a radula. The eyes are never more than two, either placed on the tentacles, or more frequently sessile at their base; but they are absent in Scaphopoda and Chitonidæ. There is a distinct organ of hearing, consisting of two round vesicles containing ciliated otolites, remarkable for their oscillatory action in living or recently killed animals.

The shell is rarely hidden in the mantle, and except in Chitonidæ it is invariably single, and most frequently spiral.

Heteropoda are sometimes included in this class. Scaphopoda are placed by Häckel and Huxley with Pteropoda; but Claus ranks them as one of the four classes of Mollusca.

There are said to be upwards of 22,000 species in this class.

Head rudimentary .... SCAPHOPODA.
Head distinct.
    Respiration by branchiæ.
        Branchiæ arborescent .... OPISTHOBRANCHIATA.
        Branchiæ pectinate .... PROSOBRANCHIATA.
    Respiration by a pulmonary sac........... PULMONIFERA.

Order I. SCAPHOPODA.

SOLENOCONCHIÆ. PROSOPOCEPHALA. CIRROBRANCHIA.

Head rudimentary; a pair of horny jaws; mouth surrounded by many filiform tentacles. No eyes. Diœcious.

There is no heart nor branchiæ. The foot is three-lobed. The shell is slender, conical, curved, and perforate at the apex, and the aperture is round. The animal is attached to the shell at the hinder part. The young are free-swimming, propelled by vibratile cilia.

There is only one family, containing, according to Schmarda, about 50 living and 125 fossil species; they mostly occur in the Devonian formation.

Dentaliidæ (Tooth-shells).

Dentalium.
Entalis = Antalis.
Gadila.

Order II. OPISTHOBRANCHIATA.

Respiration aquatic, effected by arborescent or fasciculate branchiæ, more or less exposed, rarely absent; and placed posteriorly. With or without a shell in the adult. Hermaphrodite. Larva with a velum and shell.

The shell is rudimentary or absent; but in many cases it is well-formed, and is always enclosed in the mantle. Tentacles are generally present, and are mostly non-retractile. The branchiæ are either covered or uncovered; in the latter case they are occasionally capable of being withdrawn into one or more branchial cavities. The nervous system is well developed; and they have, with few exceptions, two eyes, which are placed behind the tentacles.

Most of the species are littoral, swimming or creeping about amongst sea-weed, but a few are pelagic. They are carnivorous. One of the commonest species on our shores is known as the sea-hare (*Aplysia depilans*); it is said to emit a violet-coloured fluid when molested, due to the presence of iodine. "Sea-slugs" seem to be a general term applied to several species or even genera. The nearest external approach to slugs is perhaps best shown in *Æolis* and the Abranchia.

Certain worm-like marine animals forming the genus *Neomenia* (= *Solenopus*) compose the "order" Telobranchiata of Koren and Danielssen.

M

Claus divides this order into two sections—Dermatobranchia
[or Gymnobranchia] and Pleurobranchia. The last contains the
Tectibranchiata and Inferobranchiata ; the first the remainder of
the following suborders:—

No shell in the adult.
   Without branchiæ .... ABRANCHIATA.
   With branchiæ.
      Branchiæ on the back .... NUDIBRANCHIATA.
      Branchiæ on both sides .... INFEROBRANCHIATA.
A shell in most ; branchiæ on one side ...... TECTIBRANCHIATA.

ABRANCHIATA (= Apneusta = Dermatopnoa).—No branchiæ ;
upper surface of the body ciliated. A shell only in the larval
state.
   A subclass for Schmarda ; but not differentiated from the Nudibranchiata by Claus.

Limapontiidæ.
Limapontia = Pontolimax.
Cenia.
Actæonia.
Rhodope.

Phyllirhoidæ.
Phyllirhoë.
Acura.

Elysiidæ.
Elysia = Actæon.

NUDIBRANCHIATA (= Notobranchia).—Branchiæ placed on the
back, often retractile. Shell only in the larval state.
   Branchiæ either cylindrical, fusiform, or club-shaped (Cera-
tobranchia); leaf-shaped, feathered, or branched (Cladobranchia);
or placed behind and arborescent (Pygobranchia).

Hermæidæ.
Proctonotus.
Antiopa.
Fiona.
Alderia.
Hermæa.

Æolididæ.
Glaucus.
Cratena.

Favorinus.
Embletonia.
Æolis.
Tergipes.

Dendronotidæ.
Doto.
Meliboea.
Hero.
Dendronotus.

Tethyididæ.
Scyllæa.
Tritonia.
Tethys.

Dorididæ.
Triopa.
Onchidoris.
Polycera.
Doris (Sea-lemon).

INFEROBRANCHIATA (= Hypobranchia; Dipleurobranchia).—
Branchiæ leaf-shaped, situated in a fold on each side. No shell.
   A family of Pleurobranchia for Schmarda, the true Pleuro-

branchia forming the second family, Monopleurobranchiata (or Pomatobranchiata).

Phyllidiidæ.	*Pleurophyllidiidæ.*
Phyllidia.	Pleurophyllidia = Diphyllidia.

TECTIBRANCHIATA (= Pleurobranchia). — Branchiæ feather-shaped, on the right, rarely on both sides. A shell in most, enclosed in the mantle.

Branchiæ lie under the edge of the mantle. Shell mostly internal, unsymmetrical, either discoidal, subspiral, or involute with one or more whorls.

Runcinidæ.	Lophocercus = Oxy-	*Tornatellidæ.*
Runcina.	noë.	Acera.
	Lobiger.	Philine.
Pleurobranchidæ.	*Aplysiidæ.*	Chelidonura.
		Bulla.
Umbrella.	Notarchus.	Cylichna.
Pleurobranchus.	Dolabella.	Aplustrum.
Pleurobranchæa.	Aplysia (Sea-hare).	Tornatella.

Order III. PROSOBRANCHIATA.

Respiration aquatic, effected by pectinate or plumose branchiæ placed in a cavity in front of the heart. Animal enclosed in a univalve shell [multivalve in Chitonidæ]. Diœcious. Larva with a velum and shell.

The shell is sometimes simple, but most commonly it is spiral; in the latter the mouth [peritreme] is variously shaped, and the whorls coil round a pillar [columella], frequently perforated by a canal [umbilicus]. The animal is always capable of withdrawing entirely within the shell, which is sometimes closed by an operculum. The mantle is placed over the back of the head, and the branchiæ are lodged beneath it, anterior to the heart. Locomotion is invariably, except in *Ianthina*, performed by crawling.

To this order belong the limpet (*Patella vulgata*), money-cowry (*Cypræa moneta*), whelk (*Buccinum undatum*), and periwinkle (*Littorina littorea*). Various species of *Purpura* and *Murex* yielded the "purple" of the ancients.

Ianthina, now generally referred to this order, is provided with a float which resembles a mass of soap-bubbles, having no organic connexion with the animal, and to which the female parent commits its ova, attached to one extremity. The float is then detached, and drifts away on the surface of the sea, the contents of the ova to be eventually warmed into a free life.

According to Carus there are 12,000 species in this order, which he distributes into upwards of fifty families. The order has also been subjected to a somewhat cumbrous set of divisions, the secondary ones dependent on characters derived from the radula, except in Siphonostomata and Holostomata, where the character is taken from the form of the mouth of the shell, produced and corresponding to the siphon of the mantle in the former, while it is rounded and entire in the latter. The table below will perhaps be sufficient to show these divisions, and the families, here admitted, that are placed under them. Claus, however, makes only two sections—Cyclobranchia and Ctenobranchia, the latter comprising Scutibranchia, Pectinibranchia, and Neurobranchia, this last being included in his "group" Tænioglossa, while Scutibranchia is conterminate with the group "Rhipidoglossa" (=Aspidobranchia).

CYCLOBRANCHIA.
    Polyplacophora Chitonidæ.
    Docoglossa ...... Patellidæ.
SCUTIBRANCHIATA ... Fissurellidæ, Haliotidæ, Trochidæ, Pleurotomariidæ, Neritidæ.
PECTINIBRANCHIATA.
  Siphonostomata.
    Tænioglossa ... Strombidæ, Doliidæ, Cypræidæ.
    Toxoglossa ...... Conidæ, Terebridæ.
    Rhachiglossa... Muricidæ, Buccinidæ, Volutidæ.
  Holostomata.
    Ptenoglossa ... Ianthinidæ, Scalariidæ, Solariidæ.
    Tænioglossa ... Cerithiidæ, Melaniidæ, Pyramidellidæ, Turritellidæ, Vermetidæ, Xenophoridæ, Naticidæ, Calyptræidæ, Littorinidæ, Paludinidæ.
NEUROBRANCHIATA (=Pulmonata operculata). Helicinidæ, Cyclostomidæ, Cyclophoridæ, Aciculidæ.

Chitonidæ.
Cryptoplax=Chitonellus.
Chiton.

Patellidæ.
Acmæa.

Lepeta.
Helcion.
Patella (Limpet).

Fissurellidæ.
Scutus=Parmophorus.

Emarginula.
Fissurella.

Haliotidæ.
Haliotis (Sea-ear).

Trochidæ.
Stomatia.

Delphinula.
Phasianella.
Trochus.
Gibbula.
Astralium = Imperator.
Margarita.
Turbo.

Pleurotomariidæ.
*Pleurotomaria.

Neritidæ.
Nerita.
Neritina.
Navicella.
*Pileolus.

Ianthinidæ.
Ianthina.
Recluzia.

Solariidæ.
*Cyclogyra.
Solarium.

Scalariidæ.
Scalaria (Wentletrap).

Volutidæ.
Marginella.
Voluta.
Cymbium.
Mitra.
Harpa.
Oliva.
Ancillaria.

Muricidæ.
Pyrula.
Columbella.
Fusus.

Latirus.
Trophon.
Typhis.
Murex.

Buccinidæ.
Buccinum (Whelk).
Nassa.
Eburna.
Purpura.
Concholepas.
Coralliophila.
Ricinula.
Cuma.
Magilus.

Conidæ.
Conus (Cone).

Terebridæ.
Cancellaria.
Pleurotoma.
Terebra.

Cypræidæ.
Cypræa (Cowry).
Ovula.
Pedicularia.

Doliidæ.
Sycotypus = Ficus.
Tritonium = Triton.
Ranella.
Dolium.
Oniscia.
Cassis.

Strombidæ.
Terebellum.
Aporrhais.
Rostellaria.
Pterocera.
Strombus.

Cerithiidæ.
Cerithium.
Potamides.

Melaniidæ.
Pirena.
Melania.

Pyramidellidæ.
Obeliscus.
Chemnitzia.
Stylifer.
Eulima.
Pyramidella.

Turritellidæ.
Turritella.

Vermetidæ.
Vermetus.
Siliquaria.

Xenophoridæ.
Xenophorus = Phorus.

Naticidæ.
Velutina.
Sigaretus.
Natica.

———

Entoconcha.

Calyptræidæ.
Pileopsis = Capulus.
Calyptræa.
Trochita.
Crepidula.
Hipponyx.
Amathina.

———

Marsenia.

Littorinidæ.

Rissoa.
Hydrobia.
Assiminia.
Littorina (Peri-
  winkle).

Paludinidæ.

Paludina.
Ampullaria.
Valvata.

Bithynia.

Helicinidæ.

Stoastoma.
Helicina.
Proserpina.

Cyclostomidæ.

Cyclostoma.
Choanopoma.
Leptopoma.
Realia.

Cyclophoridæ.

Pomatius.
Cyclotus.
Cyclophorus.
Diplommatina.
Pupina.

Aciculidæ

Truncatella.
Acicula.

Order IV. PULMONIFERA.

PULMONATA. PULMOGASTEROPODA.

Respiration either aërial or aquatic, carried on by the lining membrane of a pulmonary cavity. Head distinct. Hermaphrodite. Larva with the velum absent or inconspicuous.

The pulmonary sac or cavity is placed near the neck on the right side, and communicates with the external air by a single aperture. Many of the terrestrial species have cutaneous folicular glands which secrete a granular mucus. A curious organ is a pyriform muscular sac, containing one or two slender conical styles, which can be thrust out through the aperture of the sac; they are found in certain snails, and with them they pierce each other's skin. They are known as "love-darts."

In some species the ova are of very large size. *Gadinia, Siphonaria*, &c. have simple patelliform shells. Some have no shells, or the shell is very small or is hidden within the mantle. The common garden-snail (*Helix aspersa*) and the slugs, of which we have five or six species belonging to *Arion* and *Limax*, are members of this order. *Amphibola* is the only genus with an operculum; it is also without tentacles.

Many species hibernate; during hibernation the shell is closed by an epiphragm, a calcareous secretion of the foot, which is dropped when hibernation is over.

Helicidæ [= Geophila; Stylommatophora] is the only exclusively land family. They have their eyes, except in *Onchidium*, placed on retractile stalks or tentacles (two or four). The remaining families [included under the name Basommatophora] are found in fresh or brackish waters. About 6500 species are contained in this order.

Auriculidæ.

Carychium.
Auricula.
Pythia.
Scarabus.
Melampus.

Limnæidæ.

Physa.
Chilina.
Amphipeplea.
Limnæa (Mud-snail).
Planorbis (Pond-snail).
Ancylus.

Gadiniidæ.

Gadinia.

Amphibolidæ.

Amphibola = Ampullacera.

Siphonariidæ.

Siphonaria.

Helicidæ.

Succinea.
Simpulopsis.
Janella.
Clausilia.
Zua.
Balea.
Azeca.
Pupa.
Achatinella.
Cylindrella.

Achatina.
Bulimus.
Helix (Snail).
Arion (Slug).
Geomalacus.
Sagda.
Hyalina.
Zonites.
Nanina.
Vitrina.
Parmacella.
Limax (Slug).
Glandina.
Ennea.
Streptaxis.
Testacella.
Onchidium.
Peronia.

Class V. HETEROPODA.

NUCLEOBRANCHIATA.

Free-swimming pelagic mollusks, mostly with a shell, and provided with a fin-like tail or with a ventral fin. With or without branchiæ. Sexes distinct.

The locomotive organs are modifications of the foot; these are directed upwards, the animal swimming on its back. In *Atlanta* with a spiral shell, the foot is provided with an operculum. *Firola* has no shell; *Carinaria* has a small conical transparent shell which hangs below, protecting the heart, branchiæ, &c. The remainder of the animal is a gelatinous transparent sac with two eyes and a pair of tentacles. *Bellerophon* is one of the genera of shells which goes to form the limestone beds of the Silurian and Carboniferous epochs.

This class is generally regarded as a simple order of Gastropoda; but it is a subclass for Claus. Its peculiarities are only developed in the later stages of the embryo.

Only two families are included in the Heteropoda.

Firolidæ.

Firola = Pterotrachea.
Carinaria.
Cardiopoda.

Atlantidæ.

Atlanta.
Oxygyrus.
*Bellerophon.

Class VI. CEPHALOPODA.

Free oceanic mollusks, with a distinct head; swimming or moving by long arms [modifications of the foot] placed around the mouth; the arms often provided with suckers [acetabula]. Mouth with two long beak-like jaws, and provided with a radula. Body sacciform. Sexes distinct.

The most highly organized of the Mollusca; the blood, however, is colourless, or with a slightly bluish, greenish, or yellowish tint, due to the presence of copper. The heart consists of a ventricle, two or four auricles, and two aortæ. Respiration is effected by two or four branchiæ placed within the mantle; the water entering into the branchial cavity is ejected through a peculiar tube or funnel [infundibulum], corresponding to the epipodium of the Pteropoda. The cerebral ganglia are often protected by a cartilage or rudimentary cranium, at the base of which are the two organs of hearing [otocysts]. The eyes approach the vertebrate type; they include a cornea, sclerotica, crystalline lens, retina, and vitreous and aqueous humours, and they are moved by muscles which arise from the orbital cartilages. In the *Nautilus* the eye is "a mere cup lined by the retina."

The change of colour noticeable in these animals in a living state is due to small pigment-cells [chromatophores] capable of dilatation, and, when contracted, appearing as so many specks.

Reproduction is by ova of comparatively large size; there is no metamorphosis, and there is no lower form of invertebrate life manifested by the embryo at any time during its development; on the other hand, "no arrested stage of development of any higher animal would produce any thing like a Cephalopod." [*Owen.*]

This class, according to Schmarda, includes 218 living and 1780 extinct forms; it is divided into two orders:—

Arms with suckers .... DIBRANCHIATA.
Arms without suckers..................... TETRABRANCHIATA.

Order I. TETRABRANCHIATA.

TENTACULIFERA.

Four branchiæ. Arms more than ten, without suckers. Shell external, camerated. No ink-bag.

The shell is divided into chambers by transverse septa, through which runs a slender tube [siphuncle], which may be either central, dorsal, or lateral. In the adult of the pearly nautilus (*Nautilus pompilius*) only one fourth of the shell is occupied by the animal, the remainder having been progressively evacuated as it required more room. The empty chambers are said to contain air. There are no important differences between the sexes. *Nautilus* is the only existent genus. The extinct forms were numerous in the Palæozoic and Secondary periods. The Ammonitidæ were almost entirely confined to the latter period.

Ammonitidæ.	*Orthoceratidæ.*	*Nautilidæ.*
*Ammonites.	*Orthoceras.	*Clymenia.
*Baculites.	*Cyrtoceras.	Nautilus.
*Ancyloceras.		

Order II. DIBRANCHIATA.

ACETABULIFERA.

Two branchiæ. Arms eight or ten, with suckers. With or without an external shell. Provided with an ink-bag.

The female *Argonauta*, and not the male, is provided with an external shell. It is secreted by the dilated dorsal pair of arms, and is not divided into chambers. *Spirula* has a chambered shell enclosed by the mantle; only a small portion of the body of the animal occupies the last chamber. Three species are known. In the extinct Liassic genus *Belemnites* the shell was conical and internal, having a "camerated and siphoniferous structure." In the remaining genera there is an internal shell in the form of a dorsal plate. which is either calcareous (as in *Sepia*) or corneous (as in *Loligo*). An ink-bag, enclosed in the visceral cavity, for the secretion of a black fluid, which is discharged through the infundibulum, is always present, even in the extinct species.

The "hectocotylus" is a peculiar modification of one of the arms of the male as a sex-organ; in performing its genetic function it becomes detached, but another is subsequently developed. It was first described, when so detached, as a parasitic worm, which Cuvier afterwards named "*Hectocotylus*."

Some species of this class attain a large size; *Architeuthis dux* is said to have arms forty feet long. They are all predatory and carnivorous animals.

The order has been divided into two suborders—Decapoda and Octopoda; and the former again divided and subdivided. The extinct forms are only known in the Secondary and Tertiary epochs.

DECAPODA.

Spirulidæ.
Spirula.

Belemnitidæ.
*Belemnites.
*Spirulites.

Sepiidæ.
Sepia (Cuttle-fish).
*Belosepia.

Sepiolidæ.
Sepiola.
Rossia.

Loliginidæ.
Loligo = Teuthis (Calamary, Squid).
Sepioteuthis.

Chiroteuthidæ.
Chiroteuthis.
Histioteuthis.
Octopodoteuthis.
Architeuthis.
Ommastrephes.
Onychia.
Enoploteuthis.

Loligopsidæ.
Loligopsis.

Cranchiidæ.
Cranchia.

OCTOPODA.

Octopodidæ.
Octopus.
Eledone.
Cirroteuthis.

Argonautidæ.
Argonauta (Paper Nautilus).
Philonexis.

Class VII. TUNICATA.

SACCOPHORA. ASCOZOA. PROTOVERTEBRATA. UROCHORDA.

Marine, simple or compound, unsymmetrical animals, protected by a coriaceous sac, or, in the compound, jelly-like skin, with two apertures [oral and atrial]. No foot. Heart a simple tube. Mostly hermaphrodite. Young tadpole-like, free-swimming.

The nervous system is confined to a single ganglion with its branches. There are no eyes, but one or more pigment-spots have been detected. There are also a liver, stomach, and convoluted intestine, the former sometimes rudimentary, but usually very large; in the latter the flexure is hæmal, according to Huxley, not neural as in Polyzoa.

The Tunicata follow the Mollusca, and are a "typus" or subkingdom for Claus. They form Schmarda's twenty-third class, which he places between Polyzoa and Mollusca. Häckel classes them with Vermes, but having a "true blood-relationship" with Vertebrata. Huxley considers them "in many respects" an isolated group.

There are two orders; but their limits are not very definite, *Pyrosoma* and *Doliolum* being in some respects intermediate.

A single ribbon-shaped branchia .... BIPHORA.
A pharynx acting as branchiæ .... ASCIDIOIDA.

Order I. BIPHORA.

Outer and inner integuments united throughout. Branchia ribbon-shaped. An opening at each extremity. Free-swimming. Sexes distinct.

Salpidæ were said to be solitary in one generation, which, by gemmation, gave birth to a connected group in the next. This was the "simplest form" of the alternations of generations. But it is now maintained by Brooks and Todare that the solitary and the grouped individuals are the offspring of the same parent, the former being the result of sexual reproduction (the female), the latter of budding (the chain of males). Like the Pyrosomatidæ, they are brilliantly luminous.

As individuals these animals are transparent, tubular in shape, and when united form a long chain, sometimes called by seamen "sea-serpents." On the west coast of Scotland they are occasionally found in vast numbers; at such times M'Intosh compares the appearance of the sea to boiled sago, and Huxley speaks of their masses through "which the voyager in the great ocean sails day after day."

Doliolidæ are pelagic, and are represented by sexual and sexless forms. They are transparent cask-shaped organisms, progressing by contracting and forcing the water out at one or the other extremity. The branchiæ consist of ciliated tubular bars, dividing the respiratory sac into two chambers.

Schmarda refers Appendiculariidæ and Pyrosomatidæ to this order.

Salpidæ.	*Doliolidæ.*
Salpa=Thalia.	Doliolum.

Order II. ASCIDIOIDA. (Ascidians.)

A dilated pharynx performing the functions of branchiæ. Outer and inner integuments only united at their apertures, or by blood-vessels at a few other points.

The pharynx acts as a respiratory organ as it passes the sea-water and nutrient matter to the stomach; it is "always exceedingly dilated," and the sides more or less perforated. The entry to this pharyngeal or "branchial" sac is occasionally guarded by

a circle of short tentacles. Hancock considered the sac to be
the rudiments of the lamellibranchiate gills. A peculiarity of
the Ascidians is the longitudinal fold in the pharynx, termed the
"endostyle;" its use is unknown, but it may be a sensitive
organ.

The outer integument, or sac, of the Ascidians "secretes upon
its surface a case, or 'test,' which may vary in consistence from
jelly to hard leather or horn:" the test is "rendered solid by
impregnation with a substance identical in all respects with the
'cellulose' which is the proximate principle of woody fibre and
forms the chief part of the skeleton of plants." An inner integument or tunic is composed of longitudinal and circular muscular fibres. Of the two apertures, the atrial leads into a large
cavity lined by a membrane or third tunic. The anus opens
near the mouth.

With a somewhat complex organization, the only vital action
seems to be an occasional ejection of water from the two openings
followed by a sudden contraction.

Reproduction, whether by buds or by ova, is of the most complex description. Some larval forms develop zooids, and these
may be "supplanted" by other zooids. The morphological conditions are peculiar and deceptive, and varied or modified according to the species. The young from the ova are like tad-
poles, swimming about by means of a long tail; but in the compound Ascidians the tail is lost before leaving the egg. As adults
they are mostly fixed to some foreign body. The compound
species may be seen coating the under surfaces of rocks on most
of our shores. Their food, according to Hancock, is "extracted
from sedimentary matters."

The Ascidians are regarded by Kowalewsky and others as the
nearest relations of the Vertebrata, in consequence of the presence in the larval form of the rod-like body (notochord), which
disappears in the mature animal, and in the Vertebrata is replaced by the spinal column. One of many objections to any
identity of development in the Ascidians and Vertebrates in this
respect is that the rod-like body is ventral in the former and is
developed in the same cavity as the viscera.

Huxley remarks that "In the Ascidians the central nervous
system is produced by the invagination of the epiblast, as in the
Vertebrata, and that, in most, the mesoblast of a caudal prolongation gives rise to an axial column flanked by paired myotomes,
which are comparable to the notochord and myotomes of the
vertebrate embryo." Other authorities, however, consider that
the Annelids stand nearer to the Vertebrata.

Appendiculariidæ [="suborder Copelatæ"] appear to retain a tadpole-like larval form through life. Of *Appendicularia flagellum*, Huxley says it has "an ovoid or flask-like body, one sixth to one fourth of an inch in length, to which is attached a long curved lanceolate appendage or tail, by whose powerful vibratory motions it is rapidly propelled through the water." It has the power of excreting from the surface, "with extreme rapidity, a mucilaginous cuticular investment, in the interior of which, as in a spacious case, the whole body is lodged."

The two British species of Pelonaiidæ are found in the mud in deep water; they are not fixed, and their outer and inner integuments are united. According to Goodsir and Forbes, "they indicate a relation to the cirrhograde Echinodermata." Clavellinidæ are social, connected by creeping prolongations of a common stem, and having a common circulation. Botryllidæ are compound, each individual imbedded in a jelly-like substance; and Ascidiidæ are simple forms, always fixed. *Hypobythius* occurs at a depth of 2900 fathoms.

Pyrosomatidæ [="suborder Luciæ"] are free-swimming aggregations of individuals united together in the form of a hollow cylinder, sometimes 14 inches long, one end of each individual opening into the cavity of the cylinder. They are highly luminous, two phosphorescent organs being found in each individual; they are pelagic, and seem merely to float about in the water.

Appendiculariidæ.
Appendicularia.

Pelonaiidæ.
Pelonaia.

Ascidiidæ.
Engyra.
Molgula.
Cynthia.
Styela.

Ciona.
Hypobythius.
Ascidia=Phallusia.
Boltenia.
Chelyosoma.
Chevreulia=
   Schizascus.

Clavellinidæ.
Clavellina.
Perophora.

Botryllidæ.
Polyclinum.
Amauroecium.
Aplidium.
Sidnyum.
Diazona.
Didemnum.
Distomus.
Botryllus.

———

Pyrosomatidæ.
Pyrosoma.

174

Subkingdom VII. VERTEBRATA.

MYELENCEPHALA.

Red-blooded animals, with the mass of the nervous centres enclosed in a bony axis [cerebro-spinal system]. Sexes always distinct.

The cerebro-spinal system is not represented among invertebrated animals, their ganglionic system being only the homologues of the vertebrate sympathetic system. Among invertebrates there is only one general cavity, in which the viscera and circulatory organs are contained. This is the "hæmal" region. In vertebrates a "neural" region, containing the great nervous masses, is also present. In the embryo the two cavities are developed at an early period.

Another character in which the vertebrate embryo differs from all others is in the possession of a "notochord" or "chorda dorsalis;" this is found in early embryonic life before the cerebrospinal axis is complete; it is a rod-like body, the "substance of the centre of the floor" of the spinal column, by which in most cases it is replaced. An amnion and allantois, fœtal membranes, are confined to reptiles, birds, and mammals.

Vertebrates only have true teeth; these are quite distinct from bone, and belong to the dermal appendages. The jaws are invariably above and below, never on each side. The muscles are always external to the bones.

Owen divides the Vertebrata into Hæmatocrya and Hæmatotherma—cold- and warm-blooded animals respectively. Huxley recognizes three primary divisions—Mammalia, Sauropsida (comprising birds and reptiles), and Ichthyopsida (amphibia and fishes). Häckel has four "main classes"—Leptocardia or Acrania; Monorhina, lampreys only; Anamnionata (=Anallontoidea), fishes and amphibia; and Amnionata (=Allantoidea), reptiles, birds, and mammals, the last comprised in Pachycardia or Craniota (=Holocrania). Ray Lankester, adopting Leptocardia (but with the name of Cephalochorda) and Craniota, includes also Urochorda (=Tunicata) in the Vertebrata.

The five universally recognized classes are:—

Cold-blooded; heart with less than two auricles and
two ventricles.
Gills present, at least in the earlier stages.
No lungs .... PISCES.
Lungs in the adult .... AMPHIBIA.

Gills never present .... REPTILIA.
Warm-blooded; heart with two auricles and two ven-
    tricles.
Oviparous; no mammary glands .... AVES.
Viviparous; mammary glands in the female......... MAMMALIA.

Class I. PISCES. (Fishes.)

Vertebrate animals, breathing by gills throughout life, and
covered by scales or naked, the scales overlapping each other or
imbedded in the skin, or taking the form of detached tubercles or
of spines. Heart with one auricle and one ventricle. Blood cold.
Limbs in the form of fins; caudal fin vertical.

Although the blood is cold, there is no doubt that it is higher
than the surrounding temperature, especially in the spawning-
season. One of the characters of this class is to have the vertebræ concave at both ends [amphicœlous]; the cavity thus
formed is filled with the remains of the notochord. The aorta is
very generally enlarged at its junction with the ventricle [conus
or bulbus arteriosus] and is capable of rhythmical contraction.
The sound (air-bladder or swim-bladder), peculiar to fishes, is
generally present, and is often connected with the œsophagus by
an air-tube. The air in this organ varies according to the species,
and is believed to be secreted by the inner membrane; in some
species there are muscles for compressing it. It is mostly simple,
but is sometimes cellular, and is variable in form even in species
of the same genus. The air-bladder is supposed to regulate the
specific gravity of the fish; but in many good swimmers it is
absent, or it may be present in one or absent in a closely allied
congener. In *Lepidosiren* and *Ceratodus* it is lung-like, and acts
as a respiratory organ. The skin of fishes is more closely connected to the underlying flesh than in other vertebrates. In most
there is a lateral line of peculiar scales, each of which is perforated by a tube communicating with a longitudinal canal, giving
passage to a mucous secretion produced by the glands beneath
and connected with cavities in the head. The use is unknown;
according to Vogt it is a system of absorbent vessels, while Leydig
considers it to be subservient to the sense of touch.

The vertebræ are of two kinds, "characterized by the direction
of the parapophyses." They vary in number from 15 to 236, or
350 in some of the sharks; but, owing to the coalescence of some

of them, it is not easy to determine their exact number. The head contains the heart and breathing-organs as well as the brain.

The organ of hearing (it has been doubted if they hear at all) is rather complicated in fishes, although a tympanum and cochlea are absent; two osseous bodies (otolites) are generally found in the vestibule.

The teeth are not always present, e. g. pipe-fishes, sturgeons, &c. In others they are very numerous, and attached to other bones besides maxillæ and mandibles; they are shed and renewed throughout the whole of their lives.

In their earlier stages of growth many fishes are known to undergo great changes, nor is growth known to be definitely arrested at maturity. The eye, however, ceases to grow at an early stage, so that "old fishes have comparatively smaller eyes than young ones." Fishes are very prolific; the roe of the cod is said to contain nine millions of ova.

A few species are viviparous. In no fish is there a trace of an amnion or of an allantois.

Fish swimming near the surface respire more oxygen than those living at greater depths; hence they die soon after they are taken out of the water, such are pilchards, salmon, mackerel, &c. Many are enabled to live in mud when hardened and dried up by the sun, and others may be kept for a time in a frozen state without destroying life.

Among the many deep-sea fishes discovered during the 'Challenger' expedition, one, *Halosaurus rostratus*, was found to live at a depth of 2750 fathoms.

Fishes were most abundant during the Old Red Sandstone epoch; the earliest traces of them are found in the Silurian rocks. Of the recent species some 12,000 are described; of these only about 220 inhabit the English seas and rivers, but there are no seas which yield so many for the table, whether in species or individuals.

It is difficult to find two writers agreeing upon the classification of fishes. Some consider that the class should be so limited as to exclude the Pharyngobranchii, the Marsipobranchii, and the Dipnoi, each of these also forming a class. Dr. Günther, taking the three groups as subclasses, and uniting Chondropterygii to the Ganoidei under the name of Palæichthyes, makes, together with the Teleostei, five subclasses. Prof. Huxley more recently adheres to the six orders adopted by him some years ago, though he has since proposed certain primary divisions in reference principally to the mode of attachment of the jaws to the skull. Another classification, founded almost exclusively on the skeleton, has been proposed by Dr. Cope; it is of a most radical character, what nearly corresponds to the Teleostei being divided into 24 orders, all with new names. A good account, by Dr. Th. Gill, of the various systems previous to 1873 is given in the 'American Naturalist,' vii. pp. 71 *et seq.*

Schmarda (1878) has six subclasses with fourteen orders. The former are Leptocardii, Cyclostomata, Selachii, " Ganoidea," Teleostei, and Dipnoi. The last four constitute the subclass Euichthyes for Claus, Leptocardii and " Cyclostomi " forming the other two subclasses of his arrangement.

The orders here adopted are those recognized by Huxley, with the addition of Cuvier's two orders Lophobranchii and Plectognathi, and Müller's order Holocephali. Günther's arrangement of the families and genera has been generally followed, except that the sequence has been reversed. Some families have only one species in each.

Without lungs, branchiæ only.
  No skull .... PHARYNGOBRANCHII.
  With a skull.
    No lower jaw............................... MARSIPOBRANCHII.
    With a lower jaw.
      With free gill-covers.
        Gills pectinate.
          Without true scales .... PLECTOGNATHI.
          With true scales.
            Scales horny, overlapping each
              other, &c. .... TELEOSTEI.
            Scales bony, not overlapping... GANOIDEI.
        Gills in tufts .... LOPHOBRANCHII.
      Gill-covers rudimentary; one branchial
        aperture .... HOLOCEPHALI.
      Gill-covers absent; 5–7 branchial aper-
        tures .... CHONDROPTERYGII.
With lungs and branchiæ....................... DIPNOI.

Order I. PHARYNGOBRANCHII.

ACEPHALA. MYELOZOA. CIRROSTOMI. ENTOMOCRANIA. ACRANIA. LEPTOCARDII. CEPHALOCHORDA.

Notochord persistent. Skeleton membrano-cartilaginous. No skull, brain, limbs, nor heart. Mouth without jaws, surrounded with cirri.

The persistent notochord extends beyond the cerebro-spinal axis, in all other vertebrates it stops behind the pituitary gland. Some of the great blood-vessels are rhythmically contractile, but the blood is colourless.   Respiration is effected by a ciliary membrane lining the pharynx, which extends to nearly half the length of the body, and is continued into a straight simple intestine. The eyes are rudimentary, there is no organ of hearing, and there are no scales.

Until very recently only one species was certainly known, the lancelet, *Amphioxus lanceolatus*; it is found on the English coast, and is probably an inhabitant of most seas.   It is a small transparent creature, with a delicate fin extending the whole length of the back and passing round the tail to the anus.   Pallas first described it as a *Limax*. Peters has a second genus from Moreton Bay, *Epigonichthys*; it has no anal nor caudal fin, and the anal aperture is median; only one species is known—*E. cultellus*.

Amphioxus=Branchiostoma.

Order II. MARSIPOBRANCHII.

CYCLOSTOMI.   MONORHINA.   DERMOPTERI.

Notochord persistent.   Skeleton cartilaginous; no ribs nor limbs.   Mouth suctorial, but without jaws.   Gills sac-like, communicating externally by six or seven holes.   No bulbus arteriosus.

The body is eel-shaped, and the skin is without scales.   The eyes are either wanting or are very small.   The mouth, longitudinal when closed, circular when open, has flexible lips capable of adhering to any smooth substance, with numerous small teeth within.   There is only one nasal opening.   In *Myxine* the lips are provided with 6–8 cirri, and its teeth are developed in the median line of the palate.

There is a distinct brain; but the skull is without sutures, and not separable from the vertebral column. There is no air-bladder. The kidneys are well developed.

The hag (*Myxine glutinosa*) is without eyes; it bores into and lives in the interior of other fish. The pride (*Ammocœtes branchialis*) is the larval form of the river-lampreys (*Petromyzon fluviatilis* and *P. planeri*); it is three years before the adult form is acquired. Conodonts, supposed to belong to the Myxinidæ, are minute Palæozoic tooth-like fossils.

There are two families.   Hyperotreta of Müller are the Myxinidæ, and his Hyperoartia are the Petromyzontidæ.

Myxinidæ.	*Petromyzontidæ.*
Myxine = Gastrobranchus (Hag).	Petromyzon (Lamprey).
Bdellostoma.	Geotria.
	Mordacia.

Order III. TELEOSTEI.

TELEOSTOMI.

Skeleton osseous; cranium of many bones, provided with a lower jaw. Gills free, pectinate, protected by a bony gill-cover [operculum], with gill-membrane and rays. Body covered with imbricated scales. Bulbus arteriosus not rhythmically contractile.

The scales are sometimes embedded in the skin, as in the Muræ-nidæ, and there are no true scales in the Siluridæ. They are, however, mostly present, thin and flexible, either with their edges entire (cycloid) or with their posterior edges toothed (ctenoid). They are marked with concentric and radiating lines. In some the scales are partially ossified, *e. g.* the Tunny. "Argentine" is the name of the silvery matter of the scale; the brighter colours are due to pigment-cells in the skin.

The skull is very complex, owing to the number of bones, which are, however, only centres of ossification. Four branchial arches on each side, articulated to the base of the cranium, support the gills. The operculum or gill-cover, attached to the hyoid arches, is composed of four flat bones, below which are the bony rays sustaining the branchiostegous or gill-membrane.

The muscular system of these fishes is made up of a series of vertical flakes (myotomes) corresponding in number with the vertebræ; they are connected together by a gelatinous tissue, which is dissolved by boiling. The vertebræ are often very numerous (236 in *Gymnotus*); they either gradually diminish in size to the end (Murænidæ), or end, as a rule, in a compressed series of anchylosed bones, from which the caudal rays proceed. A caudal fin so formed has almost always two equal lobes (homocercal).

Leptocephalus and *Hyoprorus* are probably, according to Günther, the offspring of Murænids arrested in their development in early life, yet continuing to grow without corresponding development, and never attaining the character of perfect animals. The former, indeed, is now said to be the larval form of the conger.

Nearly all the ordinary fishes (about 9000 species) are contained in this order; among them are the eel (*Anguilla acutirostris*), conger eel (*Conger vulgaris*), sprat (*Clupea sprattus*), sardine (*Clupea sarda*), pilchard (*Clupea pilchardus*), herring (*Clupea harengus*) [the young is the whitebait], shad (*Alausa communis*), anchovy, (*Engraulis encrasicholus*), loche (*Cobitis barbatula*), barbel (*Barbus vulgaris*), bream (*Abramis brama*), roach (*Leuciscus rutilus*), dace (*Leuciscus vulgaris*), chub (*Leuciscus cephalus*), bleak (*Alburnus lucidus*), minnow (*Phoxinus minimus*), gudgeon (*Gobio fluviatilis*), tench (*Tinca vulgaris*), carp (*Cyprinus carpio*), goldfish (*Cyprinus auratus*), pike (*Esox lucius*), smelt (*Osmerus eperlanus*), salmon (*Salmo salar*) [parr, smolt, salmon-peal, or grilse &c. the earlier stages, kelt is the female after spawning], trout (*Salmo fario*), sea-trout (*Salmo trutta*), charr (*Salmo salvelinus*), gwyniad (*Coregonus lavaretus*), holibut (*Hippoglossus vulgaris*), turbot (*Rhombus maximus*), brill (*Rhombus vulgaris*), sole (*Solea vulgaris*), plaice (*Pleuronectes platessa*), flounder (*Pleuronectes flesus*), ling (*Molva vulgaris*), hake (*Merluccius vulgaris*), cod (*Gadus morrhua*), whiting (*Gadus merlangus*), pollack (*Gadus pollachius*), haddock (*Gadus æglefinus*), bib or whiting-pout (*Gadus luscus*), mullet (*Mugil cephalus*), atherine or sand-smelt (*Atherina presbyter*), sword-fish (*Xiphias gladius*), scad (*Caranx trachurus*), dory (*Zeus faber*), pilot-fish (*Centronotus ductor*), tunny (*Thynnus vulgaris*), bonito (*Thynnus pelamys*), mackerel (*Scomber scombrus*), barracouta (*Sphyræna baracuda*), bream [sea] (*Pagellus erythrinus*), red mullet or striped surmullet (*Mullus surmuletus*), basse (*Labrax lupus*), and perch (*Perca fluviatilis*).

The Teleostei are sometimes ranked as a subclass, and four of the five divisions below as orders, the Apoda and the Abdominalia forming the Physostomi; and to these are sometimes added the Plectognathi and the Lophobranchii, which are here treated as separate orders. The Siluridæ, as supposed descendants of the sturgeons, are placed by Cope after Amiidæ. *Pegasus* is an anomalous form, generally placed with the pipe-fishes.

Fin-rays soft, articulate.
    Ventral fins none .... APODA.
    Ventral fins abdominal.................... ABDOMINALIA.
    Ventral fins pectoral or jugular .... ANACANTHINI.
    Rays of the dorsal, ventral, and anal fins
      spinous .... ACANTHOPTERYGII.

APODA.—Fin-rays soft; no ventral fins. Swim-bladder furnished with an air-tube.

Voracious fresh- and salt-water fish. Body long, slender; the scales are minute and embedded in the skin. *Gymnotus* has two sets of electrical organs.

Murænidæ.	Ophichthys.	*Gymnotidæ.*
Saccopharynx.	Muræna.	Carapus.
Anguilla (Eel).		Sternarchus.
Conger.	*Symbranchidæ.*	Rhamphicthys.
Ophisurus.	Symbranchus.	Gymnotus (Electric
Myrus.	Monopterus.	eel).

ABDOMINALIA.—Fin-rays soft; the first dorsal and first pectoral rays frequently spinous. Ventral fins behind the pectoral. Swim-bladder furnished with an air-tube.

Fresh- and salt-water fish; a few only, chiefly Cyprinidæ, are vegetable feeders. The skin is either naked or covered with cycloid scales; some of the Siluridæ are encased in bony plates; Salmonidæ have the second dorsal fin without rays. Amblyopsidæ are remarkable for having the anus under the throat, hence they have received the name of Heteropygii. In *Anableps* the iris is perforated by two pupils. *Malapterurus* is electrical. The male *Arius* carries the eggs in its mouth until they are hatched. One species, the flying-fish, is capable of sustaining itself out of the water for about 300 yards, turning round and rising and falling with the swell of the sea.

Claus has nine families in this group. It forms the order Malacoptera of Schmarda.

Halosauridæ.	Alausa (Shad).	*Cyprinidæ.*
Halosaurus.	Clupea (Pilchard, &c.).	Cobitis (Loche).
Notopteridæ.	Engraulis(Anchovy).	Barbus (Barbel). Alburnus (Bleak).
Notopterus.	*Osteoglossidæ.*	Abramis (Bream, or Carp-bream).
Alepocephalidæ.	Osteoglossum.	Leuciscus (Roach, Dace, &c.).
Alepocephalus.	Sudis (Piraracu).	Phoxinus (Minnow).
Chirocentridæ.	*Hyodontidæ.*	Gobio (Gudgeon). Tinca (Tench).
Chirocentrus.	Hyodon.	Labeo.
Clupeidæ.	*Gonorhynchidæ.*	Osteochilus. Cyprinus (Carp).
Elops. Pristigaster.	Gonorhynchus.	Catastomus.

Amblyopsidæ.
Amblyopsis.

Cyprinodontidæ.
Pœcilia.
Anableps.
Fundulus.
Protistius.
Orestius.
Cyprinodon.

Scomberesocidæ.
Exocœtus (Flying-fish).
Hemirhamphus.
Scombresox.
Belone (Garpike).

Umbridæ.
Umbra.

Esocidæ.
Esox (Pike).

Gymnarchidæ.
Gymnarchus.

Mormyridæ.
Mormyrus.

Galaxiidæ.
Galaxias.

Percopsidæ.
Percopsis.

Bathythrissidæ.
Bathythrissa.

Salmonidæ.
Salanx.
Thymallus.
Argentina.
Coregonus.
Osmerus (Smelt).
Salmo (Salmon, &c.).

Stomiatidæ.
Stomias.

Scopelidæ.
Alepidosaurus.
Paralepis.
Scopelus.
Aulopus.
Saurus.

Sternoptychidæ.
Chauliodus.
Sternoptyx.

Haplochitonidæ.
Haplochiton.
Prototroctes.

Characinidæ.
Myletes.
Serrasalmo.
Gastropelecus.
Tetragonopterus.
Alestes.
Anostomus.
Curimatus.
Erythrinus.

Siluridæ(Cat-fishes).
Aspredo.
Loricaria.
Hypostomus.
Malapterurus.
Doras.
Arius.
Pimelodus.
Amiurus.
Macrones.
Silurus (Sheat-fish).
Clarias.

ANACANTHINI.—Fins without spinous rays. Ventral fins, when present, either pectoral or abdominal. Scales cycloid or ctenoid. Air-bladder not always present.

Salt-water fish, mostly very voracious. Pleuronectidæ have unequally developed sides, the upper coloured, the lower white; the eyes are normally placed in the young, but one of them passes gradually to the other or upper side; they are either on the right side or the left according to the species. They have been separated as an order under the name of Heterosomata. Of the remaining families—the Anacanthini Gadoidei of Günther—nothing can be said collectively, except that their scales are very small or wanting. The burbot (*Lota vulgaris*) and the "freshwater trout" of Australia and Tasmania (*Gadopsis marmoratus*) are river-fish.

Pleuronectidæ.

Cynoglossus.
Hippoglossus (Holibut).
Rhombus (Turbot).
Synaptura.
Solea (Sole).
Pleuronectes (Plaice, &c.).
Arnoglossus (Scaldfish).

Ateleopodidæ.

Ateleopus.

Macruridæ.

Coryphænodes.
Macrurus.

Ophidiidæ.

Oxybeles = Fierasfer.
Ophidium.
Ammodytes (Sandlaunce, or Sandeel).
Brotula.

Gadidæ.

Molva (Ling).

Lota (Burbot).
Motella (Rockling).
Brosmius (Tusk).
Raniceps.
Chiasmodus.
Phycis.
Merluccius (Hake).
Gadus (Cod, &c.).

Lycodidæ.

Lycodes.
Gymnelis.

Gadopsidæ.

Gadopsis.

ACANTHOPTERYGII.—Dorsal, anal, and ventral fins with spinous rays. Scales cycloid in most. Inferior pharyngeal bones sometimes united (Pharyngognathi).

Mostly salt-water fish. Pharyngognathi of some authors include Chromidæ, Labridæ, Pomacentridæ, and Scomberesocidæ, but the limits are uncertain. Embioticidæ (Holconoti) are freshwater fishes from Japan and California, whose females retain their young in a pouch at the end of the ovarium until they are one-third grown. Osphromenidæ (=Labyrinthica) have a peculiar cavity above the gills for retaining the water, whereby they are enabled to remain on land for several days at a time. Cataphracti are synonymous with Triglidæ, and Pediculata or Halibatrachi with Lophiidæ. The ventral fins in Cyclopteridæ are modified into an adhesive disk; while in the remora (*Echeneis remora*) the spinous dorsal is so modified. Chætodontidæ have the dorsal and anal fins scaly at the base. Sticklebacks are remarkable among fishes as almost the only nest-builders; the work is done by the males.

Chromidæ.

Cichla.
Heros.
Acara.
Chromis.

Gerridæ.

Gerres.

Embioticidæ.

Ditrema = Embiotica=Holconotus.
Hysterocarpus.

Labridæ.

Odax.
Scarus.

Gomphosus.
Iulis.
Epibulus.
Platyglossus.
Cossyphus.
Chœrops.
Crenilabrus.
Labrus (Wrasse).

Pomacentridæ.
Amphiprion.
Pomacentrus.
Glyphisodon.

Notacanthidæ.
Notacanthus.

Mastacembelidæ.
Rhynchobdella.
Mastacembelus.

Fistulariidæ.
Aulostoma.
Fistularia.

Centriscidæ.
Amphisile.
Centriscus.

Psychrolutidæ.
Psychrolutes.

Gobiesocidæ.
Lepadogaster.
Gobiesox.
Cotylis.
Sicyastes.

Cepolidæ.
Cepola.

Trichinotidæ.
Trichinotus.

Ophiocephalidæ.
Ophiocephalus.

Mugilidæ.
Myxus.
Agonostoma.
Mugil (Mullet).

Atherinidæ.
Tetragonurus.
Atherina (Sand-smelt).

Luciocephalidæ.
Luciocephalus.

Osphromenidæ.
Anabas ("Climbing perch").
Osphromenus.

Polycentridæ.
Polycentrus.

Nandidæ.
Plesiops.
Nandus.

Malacanthidæ.
Malacanthus.

Hoplognathidæ.
Hoplognathus.

Acronuridæ.
Ceris=Keris.
Prionurus.
Acronurus.
Acanthurus.

Teuthidæ.
Teuthis.

Lophotidæ.
Lophotes.

Trachypteridæ.
Stylophorus.
Regalecus.
Trachypterus.

Comephoridæ.
Comephorus.

Acanthoclinidæ.
Acanthoclinus.

Blenniidæ.
Anarrhicas (Wolf-fish).
Pataecus.
Centronotus=Gunnellus.
Salarias.
Zoarces.
Blennius (Blenny).

Lophiidæ.
Chironectes.
Malthe.
Chaunax.
Antennarius.
Lophius (Angler).

Batrachidæ.
Batrachus.

Oxudercidæ.
Oxuderces.

Cyclopteridæ.
Liparis.
Cyclopterus (Lump-fish).

Gobiidæ.

Callionymus.
Eleotris.
Gobius.

Xiphiidæ.

Xiphias (Sword-fish).
Histiophorus.

Carangidæ.

Cyrtus = Kurtus.
Pempheris.
Equula.
Antigonia.
Capros (Boar-fish).
Psenes.
Zanclus.
Platax.
Psettus.
Trachynotus.
Lichia.
Argyriosus = Vomer.
Caranx (Scad).

Scomberidæ.

Lampris (Opah).
Pteraclis.
Brama.
Coryphæna (Dolphin).
Stromateus.
Cyttus.
Zeus (Dory).
Cybium.
Auxis.
Echeneis (Remora).
Naucrates (Pilot-fish).
Pelamys.
Thynnus (Tunny).
Scomber (Mackerel).

Trichiuridæ.

Gempylus.
Thyrsites.
Dicrotus.
Trichiurus.
Lepidopus.

Sphyrænidæ.

Sphyræna (Barracouta).

Polynemidæ.

Polynemus (Mangofish).

Sciænidæ.

Corvina.
Otolithus.
Sciæna.
Eques.
Umbrina.
Micropogon.
Pogonias.

Trachinidæ.

Opisthognathus.
Sillago.
Percis.
Trachinus (Weever).
Agnus.
Uranoscopus.
Cathetostoma.

Triglidæ.

Dactylopterus.
Peristedion = Peristethus.
Agonus.
Trigla (Gurnard).
Prionotus.

Cottus (Father-lasher).
Scorpæna.
Pelor.
Synanceia.
Pterois.
Sebastes.
Chirus.

Cirrhitidæ.

Chironemus.
Cirrhites.

Chætodontidæ.

Toxotes.
Drepane.
Ephippus.
Holacanthus.
Chelmon.
Chætodon.
Heniochus.

Sparidæ.

Chrysophrys.
Sparus.
Pagellus (Bream).
Pagrus.
Sargus.
Cantharus.

Mullidæ.

Mullus (Surmullet).
Upeneus.

Pristipomidæ.

Cæsio.
Smaris.
Mæna.
Synagris.
Dentex.
Scolopsis.
Hæmulon.

Diagramma.
Pristipoma.
Therapon.
Helotes.

Aphredoderidæ.
Aphredoderus.

Percidæ.
Dules.
Pomotis.
Bryttus.
Centrarchus.
Huro.
Arrhipis.

Grystes.
Apogon.
Ambassis.
Pentaceros.
Mesoprion.
Genyoroge.
Rhypticus (Soap-
   fish).
Plectropoma.
Serranus.
Anthias.
Centropristis.
Enoplosus.
Niphon.
Etelis.

Aspro.
Lucioperca.
Acerina.
Lates.
Labrax (Basse).
Perca (Perch).

Berycidæ.
Beryx.
Myripristis.
Holocentrum.

Gasterosteidæ.
Gasterosteus (Stickle-
   back).

Order IV. LOPHOBRANCHII.

Skeleton only partially osseous. Gills placed in tufts on the branchial arches. Body covered with plates united at their edges.

The jaws are united and tubular; and the mouth is without teeth. There are no ribs, and the air-bladder is without an air-duct. The tail is prehensile in *Hippocampus.* The fins are imperfectly developed.

The *Syngnathi* "are supposed to be able, by dilating their throat at pleasure, to draw their food up their cylindrical beak-like mouths, as water is drawn up the pipe of a syringe."

In some of the Syngnathidæ the males have a subcaudal pouch, formed by two flaps of the skin, into which the female casts her roe, and in which the young are hatched ; in others there are hemispherical depressions in which the eggs are placed. In *Hippocampus* there is also a pouch, but opening by a vertical fissure. It has been observed that when "any unusual care is taken of the eggs or young of fishes, the duty devolves upon the male." *Pegasus* has laminate gills, as in the ordinary fishes; its place is probably near Triglidæ.

Syngnathidæ.
Nerophis.
Syngnathus (Pipe-
   fish).
Stigmatophora.

Hippocampidæ.
Phyllopteryx.
Hippocampus (Sea-
   horse).
Solenognathus.

Solenostomidæ.
Solenostoma.
———

Pegasidæ.
Pegasus.

Order V. PLECTOGNATHI.

SCLERODERMI. GYMNODONTES.

Skeleton only partially osseous. Gills pectinate; a narrow gill-aperture on each side. No true scales, but either bony plates firmly united or a spiny skin, or naked.

The body is either covered by large bony plates united together to form an inflexible case, as in Ostraciontidæ, with small scale-like movable plates as in Balistidæ, with a rough granulated or spiny skin as in Gymnodontidæ, or naked as in *Orthagoriscus*. The maxillary and intermaxillary bones are united, forming the jaw, which is immovable. The gill-covers are hidden beneath the skin.

The Globe-fishes, a few of which are found in fresh water, blow themselves up by swallowing air, which is retained in a dilatation of the œsophagus. They have as well an air-bladder. Their teeth are represented by the ivory-clad termination of their jaws. They form, with *Orthagoriscus*, the family Gymnodontes of Cuvier. His Sclérodermes comprise Ostraciontidæ and Balistidæ.

Ostraciontidæ.	Monacanthus.	Diodon.
Ostracion (Trunk-fish).	Triacanthus.	Triodon.

Balistidæ (File-fish).	*Gymnodontidæ* (Globe-fish).	*Orthagoriscidæ.*
Balistes.	Tetrodon.	Orthagoriscus=Mola (Sun-fish).

Order VI. CHONDROPTERYGII.

NANTES. CARTILAGINEI. PLAGIOSTOMI. PLACOIDEI. ELASMOBRANCHII. SELACHII.

Skeleton cartilaginous; skull without sutures. Gills fixed, communicating externally by five to seven slit-like apertures. No gill-cover, or rudimentary. Bulbus arteriosus rhythmically contractile.

The mouth is transverse, placed beneath the head. The ribs are small or rudimentery. The tail is unsymmetrical (hetero-cercal). The skin is composed of small portions of dentine in the form of granules, tubercles, or spines. There is no air-bladder; and the optic nerve is not decussating. In the males are two cylindrical appendages, one on the inner edge of each ventral fin, supposed to be used as "claspers."

Many species of this order bring forth their young enclosed in horny pouches, known as sea-purses, mermaid's eggs, &c. These are secreted by a gland in the oviduct.

Among the members of this order are—the shark or white shark (*Carcharias vulgaris*), blue shark (*Carcharias glaucus*), basking shark (*Selache maxima*), dogfish (*Scyllium canicula*), thresher (*Alopias vulpes*), angel-fish (*Squatina vulgaris*), torpedo (*Torpedo marmorata*), skate (*Raia batis*), thornback (*Raia clavata*), sting-ray (*Trygon pastinaca*), eagle-ray (*Myliobatis aquila*), and saw-fish (*Pristis antiquorum*).

The Chondropterygii, notwithstanding their high development, but also accompanied by reptilian characters, are among the oldest known fish. *Cestracion Philippi* seems to have been their nearest modern representative. The fossils known as "Ichthyodorulites" are the spiny fin-rays of these fish.

There are two suborders :—

Gill-openings inferior .... BATOIDEI.
Gill-openings lateral.......................... SELACHOIDEI.

About 300 species are known, nearly equally divided between the two suborders.

BATOIDEI (Rays).—Gill-openings free, inferior. No gill-cover. No anal fin. Body broad, depressed. Notochord not persistent. Oviparous.

The pectoral fin is mostly continued to the snout. The upper lid is united to the eye, or absent. The teeth are flattened.

Torpedo has an electrical organ, composed of vertical tubes on each side between the pectoral fin and the head. *Myriosteon* of Dr. Gray, which he regarded as one of the spines of an unknown starfish, is one of the three or four hollow cylindrical tubes forming a portion of the endoskeleton of the snout of the saw-fish. According to Günther, these are modifications of the toothed processes of the cranial cartilage.

Raiidæ.	*Myliobatidæ.*	*Torpedinidæ.*
Raia (Ray, Skate).	Cephaloptera.	Temera.
*Asterodermus.	Myliobatis.	Narcine.
	Ætobatis.	Torpedo (Torpedo).
	*Rhinoptera.	
Trygonidæ.		
Pteroplatea.	*Rhinobatidæ.*	*Pristidæ.*
Trygon (Sting-ray).	Rhinobatus.	Pristis (Saw-fish).

SELACHOIDEI (Sharks).—Gill-openings mostly fine, lateral. No gill-covers. An anal fin. Body fusiform. Notochord not persistent. Ovoviviparous.

The pectoral fin is free. In some species there are two spiracles on the top of the head communicating with the pharynx. The teeth are mostly in several rows, compressed, triangular, and sometimes trilobed, with their edges serrated; they are united by ligament to the jaws.

A fœtal peculiarity is the existence of external filamentary branchiæ, as in the Batrachia; they are early removed by absorption.

Some of these fish attain to a large size—*Carcharodon Rondeletii* 40 feet; *Selache maxima* 30 feet; *Carcharias vulgaris* 25 feet [not found in the English seas]; *Alopias vulpes* 13 feet; *Lamna cornubica* 9 feet. The blue shark (*Carcharias glaucus*), 8 or 9 feet, is sometimes caught on the Cornish coast. They are nearly all very bold and voracious.

Squatinidæ.
Squatina = Rhina (Angel-fish).

Pristiophoridæ.
Pristiophorus.

Spinacidæ.
Echinorhinus.
Centrophorus.
Scymnus.
Læmargus.
Spinax.
Acanthias.
Centrina.

Cestraciontidæ.
Cestracion.
*Acrodus.
*Ptychodus.

Hybodontidæ.
*Hybodus.

Scylliidæ.
Scyllium (Dog-fish).
Pristiurus.

Notidanidæ.
Notidanus.

Rhinodontidæ.
Rhinodon.

Lamnidæ.
Selache.
Carcharodon.
Alopias (Thresher).
Lamna (Porbeagle).

Carchariidæ.
Carcharias.
Mustelus (Hound).
Zygæna = Sphyrna.
Galeus (Tope).

Order VII. HOLOCEPHALI.

Notochord persistent, skeleton cartilaginous; skull without sutures. One gill-aperture, with a rudimentary gill-cover and membrane. Skin naked.

The rudimentary gill-cover is concealed by the skin. Although there are five openings in the gill-cavity there is only one external opening. The upper jaw is united to the skull; and, in lieu of teeth, there are four hard plates above and two below.

In the male of *Callorhynchus antarcticus* there is a peculiar prehensile organ on the upper part of the snout. The subarctic and Mediterranean *Chimæra monstrosa* is known to the northern fishermen as " King of the herrings."

There is only one family :—

Chimæridæ.	
Chimæra.	*Edaphodon.
Callorhynchus.	*Passalodon.
	*Elasmodus.

Order VIII. GANOIDEI.

Skeleton variously ossified ; gills free ; gill-aperture with gill-cover. Body covered with plates or scales which are composed of two layers, the upper of enamel, the lower of bone, or occasionally naked.

The notochord is frequently persistent in the extinct forms. The vertebræ, when developed, are amphicœlous, except in *Lepidosteus*, which approaches the Reptilian type in being opisthocœlous. The bulbus arteriosus, provided with numerous valves, is rhythmically contractile, as in the Chondropterygii. The fins are often studded on the fore edge with a single or double row of spiny scales (fulcra). The tail-fin is mostly unsymmetrical.

The living species are only about 35 ; the extinct forms are numerous (about 600 species), occurring principally in the Old Red Sandstone ; among the recent ones we have the bony pike (*Lepidosteus osseus*), sturgeon (*Acipenser sturio*), beluga (*Acipenser huso*), and sterlet (*Acipenser ruthenus*). They generally ascend rivers for the purpose of spawning.

Isinglass is prepared from the air-bladder of sturgeons ; caviare or botargo is the roe (sometimes weighing 800 lbs.) of the beluga.

The Ganoids have been divided into two groups or orders (J. Müller)—Chondroganoidea or Chondrostei [a cartilaginous skeleton] and Osteoganoidea or Holostei [an osseous skeleton]. Another mode of division is into Placoganoidea [more or less protected by large osseous plates] and Lepidoganoidea [covered with scales]. The Crossopterygidæ of Huxley comprise Polypteridæ, Holoptychiidæ, &c. Schmarda's arrangement is here followed, except that Ceratodidæ are referred to Dipnoi.

(*Chondroganoidea.*)	*Pterichthyidæ.*	*Pycnodontidæ.*
Cephalaspidæ.	*Coccosteus.	*Platysomus.
*Cephalaspis.	*Pterichthys.	*Pleurolepis.
*Pteraspis.		*Pycnodus.

Acipenseridæ.
Acipenser (Stur-
geon, &c.).
Scaphirhynchus.
*Chondrosteus.

Spatulariidæ.
Spatularia = Polyo-
don.

(*Osteoganoidea.*)
Polypteridæ.
Polypterus.
Calamichthys.

Acanthodidæ.
*Acanthodes.
*Ctenacanthus.

Dipteridæ.
*Dipterus = Cteno-
dus.

Lepidotidæ.
*Lepidotus.
*Dapedius.
*Palæoniscus.

Lepidosteidæ.
Lepidosteus (Bony
Pike).

Holoptychiidæ.
*Holoptychius.
*Glyptolepis.

Cœlacanthidæ.
*Cœlacanthus.

Amiidæ.
*Leptolepis.
Amia.

Order IX. DIPNOI.

DIPNEUSTI. SIRENOIDEI. PROTOPTERI. LEPIDOTA.

Skeleton partially osseous. Gills free, a narrow gill-aperture with a rudimentary gill-cover. Air-bladder double, lung-like. Scales cycloid.

The notochord is persistent; but the cranial bones are distinct. In *Lepidosiren* the pectoral and pelvic fins are subulate and many-jointed. There is a continuous vertical fin posteriorly. Respiration is effected by the lung, a modification of the air-bladder, as well as by the gills. The heart has two auricles and one ventricle. The lung is double in Protopteridæ, single in *Ceratodus*.

Lepidosiren paradoxa and another species inhabit the Amazon; *Protopterus annectens* the Gambia, Nile, &c.; *Ceratodus* is from Queensland. The two former genera are known as "mud-fish," from the habit of burying themselves in the mud in the dry season.

The Dipnoi are a suborder of Ganoidei for Günther, while they constitute a class for others. Claus divides them into two sub-orders—Dipneumona for Protopteridæ, and Monopneumona for Ceratodidæ. Schmarda puts the Teleostei between this order and the last.

Protopteridæ.
Lepidosiren.
Protopterus = Rhino-
cryptis.

Ceratodidæ.
Ceratodus (Bar-
ramunda.)

Class II. AMPHIBIA.

Psilodermata. Nudipelliferes.

Vertebrate animals breathing by gills, mostly external in the earlier stages of life, afterwards by lungs. With or without limbs; never with fins in the adult. Heart with two auricles and one ventricle. No amnion nor allantois.

In some Amphibia the gills are retained through life, notwithstanding the presence of lungs; but much of the respiration is also carried on by the skin. They all undergo a metamorphosis, the young gradually departing from their fish-like form and developing limbs.

The lungs are generally long and narrow, sometimes extending nearly to the anus. There is no diaphragm. The kidneys are homogeneous in texture, as in fishes, reptiles, and birds. The tympanitic cavity in the ear first makes its appearance in the Batrachia. There is a single occipital condyle on each side. True ribs are wanting, or they are only rudimentary, or are not supplemented by sternal ribs. The vertebræ vary from 10 to 230. Reproduction is by ova.

None of the Amphibia are poisonous, but several secrete a very acrid fluid in their subcutaneous glands.

There are four orders:—

Recent.
    Without limbs; vermiform............... Ophiomorpha.
    With limbs; never vermiform.
      Body elongated, tailed .... Urodela.
      Body short, tailless .... Batrachia.
  Extinct .... Labyrinthodonta.

Order I. OPHIOMORPHA.

Apoda. Gymnophiona. Pseudophidia. Ophiosoma. Peromela.

Apodal vermiform amphibians, with small scales imbedded in the soft skin. Anus terminal. Eyes rudimentary or wanting. Young breathing by gills.

The body is transversely grooved or ringed. The ribs are numerous and very short, and there is no sternum. The teeth are sharp and recurved, and there is a short fleshy tongue. From the little that is known, the gills appear to be sometimes external, sometimes internal.

These animals, often several feet in length, burrow beneath the soil in tropical countries and occasionally take to the water ; they live, it is believed, on vegetable matter.

There is only one family:—

Cæciliidæ.

Siphonops.	Epicrium.
Rhinatrema.	Cæcilia.

Order II. URODELA.

GRADIENTIA. ICHTHYOMORPHA. SOZURA. CAUDATA. SAUROBATRACHIA.

Tailed amphibians with lizard-like bodies; the caudal vertebræ numerous and distinct. Two or four feet. Gills either retained through life [perennibranchiate] or disappearing at maturity [caducibranchiate].

The skin is always soft and naked, occasionally warty, and sometimes prolonged on the back into the appearance of a fin. The mouth is furnished with numerous small teeth and a short tongue. The branchial openings disappear with the gills. The ribs are short or rudimentary.

The axolotl (*Siredon*) is the larval state of *Amblystoma* ; but it sometimes remains in that state throughout life, and is at the same time most prolific, whilst those which must be supposed to have attained a higher form are utterly sterile, the sexual organs becoming apparently atrophied. From the observations of Duméril, the gills appear to be readily reproduced when lost.

The Urodela are unknown in the southern hemisphere beyond the tropic ; they are most numerous in North America. In this country we have only three species, which are known as efts or newts, the commonest being *Triton punctatus*. In the limestone grottoes of Carniola is found the curious *Proteus anguinus*. A Japanese species, *Cryptobranchus maximus*, attains a length of six feet. The mud-eels of North America belong to the genus *Siren* ; some of these attain a length of three feet.

Two divisions are usually adopted—Ichthyodea (three pairs of gills, external or internal, and amphicœlous vertebræ), and Salamandrina (without gills or gill-openings, and opisthocœlous

O

vertebræ). These include nine families, according to Claus; while Schmarda (1878), ignoring any higher divisions, has only five.

ICHTHYODEA.
Proteidæ.
Siren.
Proteus.
Menobranchus.

Cryptobranchidæ.
Amphiuma.

Menopoma.
Cryptobranchus = Sieboldia.

SALAMANDRINA.
Amblystomidæ.
Plethodon.
Pectoglossa.

Geotriton = Sperlepes.
Amblystoma.

Salamandridæ.
Triton (Eft, Newt).
Salamandra.
*Telerpeton.

Order III. LABYRINTHODONTA.

ARCHEGOSAURIA. GANOCEPHALA. STEGOCEPHALA.

Extinct amphibians, "with relatively weak limbs and a long tail." Teeth conical, their structure complex.

The body appears to have been defended by bony plates of various sizes according to the parts on which they were placed. The teeth were more or less indented by convoluted folds converging towards the centre.

The Labyrinthodonta were colossal animals of a salamandriform type, living mostly in the Triassic period. Their footprints afforded us the first indication of their having ever existed; and the unknown animal that made them, then supposed to be a kangaroo, received from Kaup the provisional name of "*Chirotherium*."

Three suborders or families are indicated—Archegosauria, Microsauria, and Mastodonsauria; one of the genera (*Archegosaurus*), it is suggested, may have been a larval form. Numerous genera have been proposed; the principal are:—

*Labyrinthodon.
*Herpeton.
*Dendrerpeton.

Order IV. BATRACHIA. (Frogs and Toads.)

BATRACHIA SALIENTIA. ANURA. THERIOMORPHA.

Tailless amphibians, breathing by lungs in the adult state.

Legs always well developed. Vertebræ procœlous. Body short and depressed. Oviparous.

In the earlier or tadpole stages the Batrachia agree with the Urodela; but eventually the tail and the gills are absorbed, the latter being replaced by two equally lobed lungs. Respiration, owing to the rudimentary ribs, is an act of swallowing.

The teeth are small, variously disposed, but sometimes absent, especially in the lower jaw. The trunk-vertebræ are few in number (7–10). There are a sternum and a pelvis. The toes are, with one exception, without claws; the hind feet are often webbed. The larynx is well developed.

The skin is shed periodically, as in the serpents; the toad swallows his. The Batrachia live on insects, small fish, or mollusks; and the bigger ones are quite capable of swallowing the smaller species.

The eggs are frequently not fecundated until after they have been laid. In *Alytes obstetricans* the female lays a chain of eggs, which the male twines round his thighs until the young leave the eggs. The female of *Pipa americana* has a soft skin on the back, in which the male embeds the eggs, which then closes over them. The female *Nototrema* has a dorsal pouch, extending over the whole of the back, in which the eggs are sheltered. Bony plates are found on the back of *Ceratophrys.*

The three British species of this order are the frog (*Rana temporaria*), toad (*Bufo vulgaris*), and natterjack (*Bufo calamita*). The little green frog of the south of Europe is *Hyla arborea*, and the edible frog is *Rana esculenta*. A Guiana frog with a most fish-like larva, attaining a large size, is *Pseudis paradoxa*. The larva of *Pipa* is tailless.

The five or six hundred species of which this order consists are very homogeneous. The subdivisions and genera [about 130] have been founded chiefly on modifications of the skeleton and disposition of the teeth. Günther (in 1858) had twenty-five families, Cope (in 1865) fourteen, Owen (in 1866) four, Mivart (in 1869) sixteen, Carus (in 1875) seventeen, and Schmarda (in 1878) five—the fifth, in addition to the four mentioned below, being represented by *Bombinator*. A higher series of divisions has been into Aglossa (tongueless) and Phaneroglossa (tongue present), the latter again into Oxydactyla (slender toes) and Platydactyla or Discodactylia (toes dilated at the tip).

The following is a tabular view of Dr. Günther's classification:—

o 2

Series.	Sections.	Families.

A. AGLOSSA.

Aglossa Haplosiphonia .... $\left\{\begin{array}{l}\text{1. Dactylethridæ.} \\ \text{2. Pipidæ.}\end{array}\right.$

Aglossa Diplosiphonia .... 1. Myobatrachidæ.

B. OPISTHOGLOSSA.

Opisthoglossa Oxydactyla.

Ranina .... $\left\{\begin{array}{l}\text{1. Ranidæ.} \\ \text{2. Cystignathidæ.} \\ \text{3. Discoglossidæ.} \\ \text{4. Asterophyidæ.} \\ \text{5. Alytidæ.} \\ \text{6. Uperoleidæ.}\end{array}\right.$

Bombinatorina .... 1. Bombinatoridæ.

Brachycephalina ...... $\left\{\begin{array}{l}\text{1. Phryniscidæ.} \\ \text{2. Brachycephalidæ.}\end{array}\right.$

Opisthoglossa Platydactyla.

Bufonina .... $\left\{\begin{array}{l}\text{1. Rhinodermatidæ.} \\ \text{2. Engystomatidæ.} \\ \text{3. Bufonidæ.}\end{array}\right.$

Hylina .... $\left\{\begin{array}{l}\text{1. Polypedatidæ.} \\ \text{2. Hylodidæ.} \\ \text{3. Hylidæ.} \\ \text{4. Pelodryidæ.} \\ \text{5. Phyllomedusidæ.}\end{array}\right.$

Micrhylina............... 1. Micrhylidæ.

Hylaplesiina .... $\left\{\begin{array}{l}\text{1. Hylædactylidæ.} \\ \text{2. Brachymeridæ.} \\ \text{3. Hylaplesiidæ.}\end{array}\right.$

C. PROTEROGLOSSA.

Rhinophrynina .... 1. Rhinophrynidæ.

The groups adopted here may be tabulated thus :—

No tongue (*Aglossa*)............................. PIPIDÆ.
With a tongue (*Phaneroglossa*).
    Toes not dilated at the end.
        With maxillary teeth ... . RANIDÆ.
        No maxillary teeth .... BUFONIDÆ.
    Toes dilated at the end .... HYLIDÆ.

Pipidæ.
Pipa.
Dactylethra.
Myobatrachus.

Ranidæ (Frogs).
Pseudis.
Rana.
Ceratophrys.
Pleurodema.
Cystignathus.
Limnodynastes.
Discoglossus.
Alytes.
Asterophrys.

Hyperolia.
Bombinator.
Pelobates.
Liopelma.
Cacotus.
Phryniscus.
Pseudophryne.
Rhinoderma.
Engystoma.

Bufonidæ (Toads).
Bufo.
Otilophus.
——
Rhinophrynus.

Hylidæ (Tree-frogs).
Acris.
Ixalus.
Polypedates.
Hylodes.
Hyperolius.
Callula.
Litoria.
Hyla.
Nototrema.
Pelodryas.
Phyllomedusa.
Micrhyla.
Hylaplesia.
Dendrophryniscus.

Class III. REPTILIA. (Reptiles.)

Vertebrate animals, breathing by lungs throughout life; the body covered with bony plates or with scales. Blood cold. With or without limbs. An amnion and allantois.

The heart has two auricles and a ventricle, but in crocodiles the latter is divided by a septum. The heart has two aortic arches. The lower jaw is attached to the skull by the intervention of an *os quadratum*, as in birds, and there is only a single occipital condyle on each side. The teeth, which are sometimes absent, are conical, and not adapted for crushing or tearing; the food is consequently swallowed entire. The teeth are not provided with fangs at the root, and they are generally anchylosed to the jaw. The ribs are always well developed, but the sternum is often wanting. The cavity of the thorax is not separated from the abdomen by a diaphragm. The segmental structure of the lateral muscles is still shown in the crocodile, but it is almost lost in the Ophidia and Chelonia.

A "purposive weapon" is found in the embryo of snakes and lizards, a sharp tooth being developed in the premaxillary bone, wherewith they cut through the egg-shell; it disappears in the adult.

There are about 1900 living species in this class; of the extinct forms over 400 are known, mostly belonging to the Secondary period. There are ten orders; Ichthyopterygia and Sauropterygia form the Enaliosauria of some writers.

Recent.
>An osseous exoskeleton.......................... CHELONIA.
>No osseous exoskeleton.
>>No sockets for the teeth.
>>>No eyelids .... OPHIDIA.
>>>With eyelids .... SAURIA.
>>Sockets for the teeth.......................... CROCODILIA.
Extinct.
>Limbs natatory.
>>No neck .... ICHTHYOPTERYGIA.
>>A long neck .... SAUROPTERYGIA.
>Limbs not natatory.
>>With molar teeth .... THERIODONTIA.
>>Without molar teeth.
>>>Teeth conical, numerous.
>>>>Four strong, unguiculate limbs ...... DINOSAURIA..
>>>>Fore limbs adapted for flying......... PTEROSAURIA.
>>>Teeth wanting, or with two tusks only ANOMODONTIA.

Order I. OPHIDIA.   (Serpents.)

Body slender, cylindrical, covered with horny scales; no visible
limbs. No eyelids. Mouth dilatable. No sacrum, sternum, or
pelvis. Vertebræ and ribs very numerous. Tongue bifid.

All the bones composing the upper and lower jaws are movably
united by ligament or muscle, and not by cartilage; the mastoid
bone is also movable, and the os quadratum often passes behind
the head. This peculiar structure allows the mouth to be enormously dilated. The vertebræ are procœlous and very numerous
[from 200 to 420]. The ribs, of which there are sometimes 300
pairs, are always free at the extremity; assisted by the ventral
scales, they are the real organs of locomotion. The teeth are slender,
curved, and anchylosed to the bones to which they belong. There
are no eyelids, but the eyes are covered by the transparent epidermis. There are no external ears. The anal cleft is transverse.

In most serpents there is one lung of great length, the other
rudimentary or aborted. The heart "agrees with other organs
in its elongate form."

In venomous serpents there are two poison-fangs firmly fixed
to the upper maxillary bones. These may be moved backwards
or forwards, so that with either movement the fangs are raised or
depressed. The fang is a long conical tooth, either grooved ex-

ternally, or it is inflected on itself so as to form a tube; the poison-gland is at its base beneath the eye on each side. The bite of a venomous serpent makes two punctured wounds, but from a non-venomous serpent there will be probably two lines of punctures. In this country we have only one venomous species, whose bite is rarely attended with fatal effects. On the other hand, in Hindostan the number of persons killed annually is estimated by Sir J. Fayrer to be over 20,000 (Thanatoph. p. 32).

The action of serpent-poison is neurotic, annihilating in some unknown way the nerve-power. Serpents, it is said, are not affected by their own poison. *Ophiophagus elaps*, the largest and most venomous of them, feeds on other serpents only less dangerous than itself. Hydrophidæ are marine, but confined to Indian seas.

The most noted serpents are the rattlesnake (*Crotalus horridus*), the asp [of Cleopatra] (*Cerastes Hasselquistii*, or *Naia haje* according to Claus), the puff-adder [of the Cape] (*Clotho arietans*), viper or adder (*Pelias berus*), cobra (*Naia tripudians*), and the snake (*Natrix torquata*). The coral-snake of Guiana is *Ilysia scytale*. The largest snakes belong to the genera *Python* and *Boa*. *Eunectes murinus* of the Amazon is over 20 feet long.

The Ophidia, of which more than 900 species have been described, are divided by Duméril and Bibron into five suborders, which may be tabulated thus:—

Mouth dilatable.
    With poison-fangs.
        Fangs perforate..................... SOLENOGLYPHA.
        Fangs grooved .... PROTEROGLYPHA.
    Without poison-fangs .... AGLYPHODONTIA.
Mouth not dilatable.
    With anal spurs .... ANGIOSTOMATA.
    Without anal spurs .... OPOTERODONTIA.

OPOTERODONTIA.—Mouth small, not dilatable. Teeth in the upper jaw only. No poison-fangs.

The tail is short or none. The head is small, and the eyes are rudimentary. There are no anal spurs.

These are small worm-like animals, sometimes referred to the lizards, living in burrows underground or under stones; they feed on insects and worms.

Typhlopidæ.
Stenostoma.
Typhlops.

ANGIOSTOMATA.—Mouth not dilatable. Teeth variable, but no poison-fangs. With anal spurs.

As in Opoterodontia, with which this suborder is sometimes combined, the mastoid, when present, is united with the temporal bone, and the os quadratum is fixed. The anal spurs are the condensed epidermis of the rudimentary hind limbs.

Cylindrophidæ.
Ilysia = Tortrix.
Cylindrophis.

Uropeltidæ.
Uropeltis.

AGLYPHODONTIA (= Colubrina = Colubriformia).—Mouth dilatable. Solid hooked teeth in both jaws. No poison-fangs.

Innocuous; a few have fangs, but no poison-glands. Rudiments of a pelvis and hind limbs are found in some of the Boidæ. Tail prehensile in many species.

Calamariidæ.
Calamaria.
Oligodon.

Colubridæ.
Coronella.
Ablabes.
Simotes.
Tachymenis.
Liophis.
Natrix = Tropido-
    notus (Snake).
Coluber.
Zamenis.
Dromicus.
Herpetodryas.

Rhachiodontidæ.
Rhachiodon.

Homalopsidæ.
Hydrops.
Herpeton.
Homalopsis.

Dryophidæ.
Dryophis.

Psammophidæ.
Psammophis.
Cœlopeltis.

Dendrophidæ.
Chrysopelea.
Dendrophis.

Dipsadidæ.
Leptodira.
Dipsas.

Scytalidæ.
Scytale.

Lycodontidæ.
Lycodon.

Boidæ.
Eryx.
Eunectes (Anaconda).
Enygrus.
Boa.
Python.
Liasis.
Morelia.
———
*Palæophis.

Acrochordidæ.
Acrochordus.

PROTEROGLYPHA.—Mouth dilatable. Fangs in the upper jaw grooved, with strong hooked teeth behind them. Poison-glands always present.

Palate and pterygoid bones with teeth, as well as the lower jaw. The tail is vertically compressed in Hydrophidæ.

Elapidæ.	Ophiophagus.	*Hydrophidæ* (Sea-
Diemenia.	Naia (Cobra).	snakes).
Bungarus.	Furina.	Hydrophis=Pelamis.
Hoplocephalus.	Elaps.	Enhydrina.
		Platurus.

SOLENOGLYPHA.—Mouth dilatable. Fangs in the upper jaw perforated. Poison-glands always present.

The head is triangular, much broader at the base than the neck. The palate and under jaw are provided with teeth. The tail is comparatively short; in the rattlesnake terminated by 20–30 horny movable rings formed by the epidermis. In *Cerastes* the epidermis is developed into small horns above the eyes.

Viperidæ.	Cerastes.	Cenchris.
Pelias (Viper or	Atractaspis.	Bothrops.
Adder).		Crotalus (Rattle-
Vipera.	*Crotalidæ.*	snake).
Clotho.	Trimeresurus.	Trigonocephalus.

Order II. SAURIA.

LACERTILIA.

Body slender, covered with horny scales; mostly with four legs. Eyes with movable lids. Mouth not dilatable.

The legs are generally four, sometimes only two, or they may be absent. In the apodal forms there are only rudiments of the shoulder-girdle, and in *Anguis fragilis* of a sternum and pelvis. The vertebræ, except in *Hatteria*, are procœlous: in the dorsal vertebræ there is only one surface on each side for the articulation of the ribs. These, except in the apodal forms, are few in number comparatively, and, in the normal forms, some of them are always attached to the sternum. The teeth are usually simple, fixed in the jaws, not in distinct sockets; the palate is rarely toothed. The upper jaw has no independent motion, as it has in the serpents, and the lower jaw forms one bone. The tympanum of the ear is nearly on a level with the skin. In some species there is a row of perforated scales on the inside of the thighs; these are the femoral pores. The anal cleft is transverse.

In most Sauria the lungs are equal, but in the blindworm (*Anguis fragilis*) the left lung is only half the length of the right. Another peculiarity of this animal is that it casts its skin like serpents.

The Sauria are oviparous or ovo-viviparous, not producing many young at a birth. They are mostly very active, enjoying the hottest sunshine, although a few are seminocturnal, and living chiefly on insects.

Out of nearly 800 species only three are found in this country, viz. two lizards (*Lacerta agilis* and *Zootoca vivipara*) and the blindworm (*Anguis fragilis*). Various large lizards of the tropics are known as Iguanas. *Ascalabotes fascicularis* is the *Tarentola* of the Italians. The glass-snake of North America is *Ophisaurus ventralis*. The Tuatara of New Zealand (*Hatteria punctata*) is the most remarkable form of the Reptilia, combining "the characters of a high and low organization."

The Sauria may be divided into eight suborders, but of these the Cyclosaura and Geissosaura are sometimes united under the name of Cionocrania or of Leptoglossæ. Gray's classification is here principally followed. Amphisbænoidæa he placed with the " Shield Reptiles," his Cataphracta.

Vertebræ procœlous.
  Tongue short, thick.
    Pupil linear .... NYCTISAURA.
    Pupil round.
      Tail short, anus terminal............... AMPHISBÆNOIDÆA.
      Tail long, anus not terminal.
        Ventral scales overlapping.
          Tongue entire or nearly so ...... STROBILOSAURA.
          Tongue notched at the tip ...... GEISSOSAURA.
        Ventral scales placed in cross bands CYCLOSAURA.
  Tongue elongate.
    Eyes covered, except round the pupil, by
      a circular lid .... DENDROSAURA.
    Eyes with two valvular lids.............. FISSILINGUIA.
Vertebræ amphicœlous......................... RHYNCHOCEPHALIA.

CYCLOSAURA (= Ptychopleuræ).—Tongue short, slightly extensible. Ventral scales arranged in cross bands.

Body either lizard- or serpent-like. There are two or four short feet or none. The back is covered with large scales.

Gray included Lacertidæ, Monitoridæ, &c. in his Cyclosaura. Claus unites this and the former in one suborder—Brevilinguia.

Chalcididæ.	*Zonuridæ.*	*Ecpleopodidæ.*
Chalcides.	Zonurus.	Ecpleopus.
Chamæsaura.	Pseudopus.	Cercosaurus.
Cricochalcis.	Ophisaurus.	
	Gerrhonotus.	

FISSILINGUIA.— Tongue slender, cleft, extensible. Eyelids present, except in *Ophiops*.

There are two movable eyelids, and sometimes a nictitating membrane. The body is never serpent-like; and the legs are always well developed. The scales are disposed in transverse bands on the belly and tail; the latter is often compressed, especially in the aquatic species. Some of the species attain a length of seven feet; *Mosasaurus*, an extinct form of the Chalk formations, of seventy-five feet.

Monitoridæ.	Ophiops.	*Xantusiidæ.*
Hydrosaurus.	Acanthodactylus.	Xantusia.
Monitor = Varanus.	Cabrita.	
Heloderma.	Tropidosaurus.	*Ameividæ.*
	Psammodromus.	Teius.
Lacertidæ.	Lacerta ⎫ (Lizard).	Ameiva.
	Zootoca ⎭	Crocodilurus.
Eremias.		
	*Mosasaurus.	

STROBILOSAURA.—Tongue short, fleshy, not extensible. Eyelids present. Ventral scales small, rhombic, overlapping. Four well-developed feet, with mostly slender unequal toes.

Body frequently compressed, and with a row of erect pointed scales on the head and back. *Draco* has the skin expanded at the side by its six false ribs, greatly elongated, by which means it is enabled to take flying leaps.

Agamidæ are acrodont, Iguanidæ pleurodont. These are lizards, frequently of large size, found on trees and on rocky ground.

Agamidæ.	Stellio.	Holbrookia.
Draco.	Hoplurus = Uromastix.	Cyclura.
Chlamydosaurus.		Iguana.
Ceratophora.	Hoplocercus = Pachycercus.	Basiliscus.
Sitana.		Anolis.
Calotes.		Polychrus.
Agama = Amphibolurus.	*Iguanidæ.*	Ecphymotes.
Trapelus.	Phrynosoma.	
	Ophryoëssa.	

NYCTISAURA (= Ascalabota).—Tongue short, fleshy, not extensible. Eyelids rarely present. Toes subequal, more or less dilated, the claws sometimes absent.

Body, and especially the head, somewhat depressed, covered with small scales more or less tuberculate. Eyes large, with frequently a linear vertical pupil.

Found under stones and in houses in most warm climates, running up the walls or on the ceilings. Many species are nocturnal.

Geccotidæ.	Phelsuma.	Hemidactylus.
Œdura.	Naultinus.	Ascalabotes.
Ptyodactylus.	Phyllurus.	Gecco.
Ptychozoon.		

DENDROSAURA (= Vermilinguia = Rhiptoglossi).—Tongue long, club-shaped, extensible. Eyes large, covered, except around the pupil, by a circular lid.

Body compressed, covered with a shagreened skin. The head is more or less crested. The toes are divided into two equal sets. The tail is round and prehensile.

Confined to the warmer parts of the Old World. They are exceeding slow in progression. The power of changing colour is common to these and to many other lizards.

Chamæleontidæ.

Chamæleon.
Rhampholeon.

RHYNCHOCEPHALIA.—Vertebræ amphicœlous. Quadrate bone immovably united to the skull.

There is a complex abdominal sternum as in the crocodiles, and the lungs have large air-cells as in the Amphibia.

This group is represented by *Hatteria punctata*, a large sluggish New-Zealand lizard, now almost extinct. A second species from Cook's Straits (*H. Güntheri*) has been described.

Hatteriidæ.

Hatteria = Sphenodon.

AMPHISBÆNOIDÆA (= Annulata).—Tongue short. Eyes very small or wanting; no eyelids. No legs, except a small fore pair in *Chirotes*.

Body cylindrical, the head and tail being alike. The scales are arranged in rings round the body.

These are serpent-like, sluggish animals living in or near ants'-nests, and are mostly nocturnal.

Amphisbænidæ.	Trogonophis.
Lepidosternon.	*Chirotidæ.*
Amphisbæna.	Chirotes.

GEISSOSAURA.—Tongue short, bifid at the tip, slightly exten sible ; ventral scales rounded, imbricate.

Body lizard-like or serpent-like, generally covered with rounded overlapping scales; the feet, when present, or more than rudimentary, are very small.

Feeble, harmless animals, living in holes or under stones, and feeding on insects and worms. *Seps* is viviparous. Our only representative of this suborder—the slowworm or blindworm—is ovo-viviparous, bringing forth from seven to twelve young at a birth. The Scinck (*Scincus officinalis*), a native of Egypt and Syria, was once highly valued for its medicinal qualities.

Scincidæ.	*Anguidæ.*	*Acontiidæ.*
Lygosoma = Hinulia	Anguis (Slowworm).	Acontias:
= Mocoa.	Diploglossus (Gally-	Typhlonurus=Ty-
Euprepes = Mabouya.	wasp).	phline.
Scincus.	Ophiodes.	*Gymnophthalmidæ.*
Cyclodus.	Ophimorus.	
Trachysaurus.		Pygopus = Bipes.
Anomalopus.	*Sepidæ.*	Aprasia.
Heteropus.	Seps = Gongylus.	Lialis.
	Scelotes.	Gymnophthalmus.
		Ablepharus.

Order III. CROCODILIA.

LORICATA. EMYDOSAURIA. HYDROSAURIA.

Teeth lodged in distinct sockets. Body covered with bony plates. Feet short, toes webbed. Ribs with bifurcate heads. Ectopterygoids present. Ventricle double.

The cervical vertebræ are remarkable for having short ribs; the dorsal vertebræ are procœlous in the recent species. A series of so-called abdominal ribs, which are not connected with the vertebræ, are formed by the ossification of tendinous portions of the rectus abdominis. There are three eyelids, a movable earlid, the only approach to an external ear in reptiles, and a short fleshy tongue. The heart has four complete chambers; but there is an opening between the two aortic arches (*foramen Panizzæ*), and consequently the mixture of venous and arterial blood does not take place in the heart itself as in other reptiles. The lungs are confined to the thorax. As in the tortoises, the anal cleft is longitudinal.

This order contains the alligator (*Alligator lucius*), the Gangetic crocodile (*Gavialis gangeticus*), and the crocodile [of the Nile] (*Crocodilus vulgaris*). "Cayman" and "Jacaré" are the names of certain Alligatoridæ given by savages and imposed on science.

Of the two extinct families, Teleosauridæ have amphicœlous, and Stenosauridæ opisthocœlous vertebræ.

There are about twenty-three recent species, all confined to fresh water. Some of the extinct forms were marine. The remains of crocodiles and alligators are found in the Tertiary deposits of England.

Teleosauridæ.	Mecistops.	Gavialis.
*Teleosaurus.	Crocodilus (Crocodile).	*Leptorhynchus.
Stenosauridæ.		*Alligatoridæ.*
*Stenosaurus.	*Gavialidæ.*	Alligator (Alligator).
Crocodilidæ.	Tomistoma = Rhynchosuchus.	Cayman. Jacaré.
Osteolæmus.		
	*Proterosaurus.	

Order IV. ICHTHYOPTERYGIA.

ICHTHYOSAURIA.

Limbs paddle-like, furnished with marginal ossicles; the digits not more than five, but with numerous phalanges. Neck very short. No sacrum nor sternum. Body fish-like.

The jaws were produced; the nostrils were placed close to the large orbits; the teeth were lodged in grooves, not in distinct sockets, and were numerous and powerful. The vertebræ were amphicœlous. A vertical caudal fin is supposed to have been present.

If we except *Eosaurus* (from the Coal-measures of North America), one genus only is known, the species of which abounded during the Mesozoic period, and it is believed were very voracious. They were supposed to be marine, coming occasionally to the shore. Coprolites are said to be their droppings.

* Ichthyosaurus.

Order V. SAUROPTERYGIA.

PLESIOSAURIA.

Limbs paddle-like, formed for swimming. Vertebræ with their articulating surfaces flat or slightly concave. Neck mostly very long. A sacrum of two bones. No sternum.

The jaws were produced, but the nostrils were placed far back near the orbits; the teeth were lodged in distinct sockets, and, except in *Placodus*, were of the normal character. The eye had no bony plates. The scapula resembled the scapula of Chelonia.

This order contains Mesozoic marine lizards, generally of large size.

*Nothosaurus.   *Plesiosaurus.   *Placodus.
*Simosaurus.

Order VI. PTEROSAURIA.

Forearm and fifth digit very long, formed for flight. Vertebræ procœlous. Jaws prolonged; teeth conical, lodged in distinct sockets.

The long bones and vertebræ had air-cells, as in birds. The neck was very long; and there was a sternum with a mesial crest. For these and other reasons, Prof. H. G. Seeley thinks them more nearly allied to birds than to reptiles; and he proposes for them a class which he names Saurornia.

These flying lizards of the Mesozoic period seemed to have been confined to mid-Europe. One appears to have had an expanse of wing of more than twenty feet.

Marsh has recently described a new form (*Pteranodon*) allied to this order found in the Cretaceous beds of Kansas. It is without teeth, and constitutes, according to him, a distinct order and a distinct family.

*Ramphorhynchus.   *Dimorphodon.   *Pterodactylus.

Order VII. DINOSAURIA.

ORNITHOSCELIDA. PACHYPODA.

Teeth lodged in distinct sockets. Vertebræ mostly flat on their articular surfaces, a few anterior opisthocœlous. Limbs ambulatory. Exoskeleton consisting of bony plates in some, in others only a naked skin.

In some of their characters the Dinosauria approach the ostriches; but their metatarsal bones were not anchylosed to the tarsus, except partially in *Compsognathus*, which it is supposed must have walked more or less in an erect position. Huxley considers that the gradation between birds and reptiles will be found in this group.

These Saurians were mostly of gigantic size; *Cetiosaurus* was 60 or 70 feet long; *Compsognathus*, however, was only about two feet long. They are found chiefly in the Oolitic and Cretaceous formations.

*Cetiosaurus.   *Megalosaurus.   *Iguanodon.
*Omosaurus.   *Compsognathus.

Order VIII. ANOMODONTIA.

DICYNODONTIA.

No teeth, or with closely-set teeth in both jaws, or two large tusks in the upper jaw. Vertebræ amphicœlous. Sacrum large. Limbs ambulatory. No exoskeleton.

In *Dicynodon* an important step towards the mammalian type is made in the constant renewal of the tusks from the matrix, as in the long-lived and ever-growing tusks and scalpriform incisors of the Mammalia. In their beak, sheathed in horn, they resembled the Chelonia. They are found in the Trias of South Africa, Europe, and Bengal.

*Oudenodon.   ——   *Dicynodon.
              *Rhynchosaurus.

Order IX. CHELONIA.

TESTUDINATA.   CATAPHRACTA.

No teeth. "Trunk-ribs broad, flat, suturally united, forming with their vertebræ, sternum, and dorsal bones an expanded thoracico-abdominal case." External nostril single. Eyelids. A sacrum.

The upper plate, or "carapace," is composed of the dorsal vertebræ and ribs; the lower, or "plastron," represents the sternum. These are formed of bony plates covered with unconformable horny scales. The vertebræ of the neck and tail are alone movable. The lungs extend into the abdominal cavity with the viscera; and, owing to the immobility of the ribs, these reptiles, like the Batrachia, swallow the air they breathe.

The Chelonia are very sluggish in their movements, very tenacious of life, and are said to pass even years without food. [*Cuvier.*] They are oviparous, and live mostly on vegetable food; Trionychidæ are carnivorous. In their jaws deprived of teeth they resemble birds. Their anal cleft is longitudinal.

This order includes the green-turtle (*Chelonia midas*), tortoise-shell or hawk's-bill turtle (*Chelonia imbricata*), logger-headed turtle (*Thalassochelys caretta*), terrapin (*Emys* sps.), box-tortoise (*Cistudo carinata*), alligator-tortoise (*Chelydra serpentina*), and soft-back (*Trionyx ferox*). The common tortoise is *Testudo græca*.

Several gigantic species of tortoise (*Testudo*), living and extinct, are known; the former are now almost entirely confined to the Galapagos and Aldabra islands, the latter to Mauritius and Rodriguez. *Colossochelys atlas*, an Indian fossil species, is said to have been 20 feet in length.

There are about 200 good species in this order, which have been distributed into 103 so-called genera, many of them, according to Günther, depending on slight differences of the skull, but unaccompanied by any external characters.

Cheloniidæ.	*Chelydidæ.*	Clemmys.
Sphargis.	Emydura.	Chelydra = Chelonura.
Thalassochelys=	Hydraspis.	
Caouana.	Podocnemis.	Emys (Terrapin).
Chelonia (Turtle).	Pelomedusa.	
	Chelys.	*Testudinidæ.*
Trionychidæ.		Teleopus.
Emyda.	*Emydidæ.*	Cinyxis.
Trionyx.	Cinosternum.	*Colossochelys.
	Cistudo.	Testudo (Tortoise).

Order X. THERIODONTIA.

Incisors defined by position, and divided from the molars by a large laniariform canine on each side of both jaws, the lower canine crossing in front of the upper. No ectopterygoids. The humerus with a supracondylar foramen.

The supra condylar foramen is one of the characters of the Felidæ. In this and certain other respects the Theriodontia had mammalian resemblances. Owen has described many genera and species whose remains have been found in the Triassic lacustrine deposits of South Africa.

*Cynodraco.   *Tigrisuchus.   *Galesaurus.

P

Class IV. AVES. (Birds.)

Vertebrate, warm-blooded, oviparous animals, breathing by lungs, and clothed with feathers. Heart with two auricles and two ventricles. No corpus callosum. An amnion and allantois.

The lower jaw is articulated to the skull by the intervention of an *os quadratum* (representing the *incus* of mammals), and the skull to the atlas by single occipital condyles. The cervical vertebræ vary from nine to twenty-three. The coracoid process of the scapula in mammals is a distinct bone in birds, and the two clavicles are united to form the furculum [merrythought]. The sternum, or breast-bone, is generally notched posteriorly [one or, more rarely, two notches on each side] or perforated. In all living birds the terminal tail-bones are anchylosed. The bones are permeated with air derived from the bronchi, but in very young birds they are filled with marrow. Air-cells are also more or less interposed between the skin and the muscles.

Most birds have an inferior larynx situated close to the bifurcation of the trachea; sound is produced here and is modified above. In singing birds it is worked by five or six pairs of muscles. The trachea is composed of bony rings, the bronchi of half-rings situated on the outer side, the inner side being completed by a membrane. There is no epiglottis, or at least it is only rudimentary; the papillæ at the base of the tongue, directed backwards, prevent the food from entering the trachea. The tongue is more an organ of touch than of taste. The œsophagus, always very dilatable, terminates in the "proventriculus," which is lined with the glands secreting the gastric juice. Close to the proventriculus is the stomach, thin and membranous in carnivorous, muscular in granivorous birds. The intestine is from three to nine times the length of the body; it is longest in graminivorous birds.

In many birds, such as those that gorge themselves at uncertain intervals, and those which, living exclusively on vegetable food, require a temporary store, the œsophagus dilates to form a "crop" [ingluvies]. From the crop, or from the stomach, many birds, such as kingfishers, parrots, pigeons, &c., have the power of producing their partially digested food for the support of the female during, and the young after, incubation; in the latter and a few others the crop secretes a peculiar milky fluid. Some birds also have the power of casting up the indigestible matter which they have swallowed.

The liver is two-lobed, rarely there is also a small central lobe. A gall-bladder may or may not be present, even in allied species, or occasionally in the same species. The kidneys are remarkable for being placed in cavities of the iliac bones, as in reptiles, and their substance is homogeneous. The lungs are attached to the ribs and spine, but they are al first free. The diaphragm is not well developed.

Birds have no eyelashes, but they possess a third eyelid [membrana nictitans], transparent or otherwise. The sclerotica anteriorly is supported by covered bony plates. There is always an ear, but no external concha. In no bird is the eye ever rudimentary, or wanting. The brain is very variable in size; it is about $\frac{1}{25}$ the weight of the body in the sparrow, and $\frac{1}{1100}$ in the ostrich. The optic thalami are small, and the optic lobes, differing from "every other class," are lateral and inferior.

The feathers in birds consist in general of the shaft [scapus], a continuation of the quill [calamus], to which the "webs" are attached on each side; these are made up of a number of barbs [rami], each furnished with hooked branches [barbules or ramuli]. The "contour-feathers" [pennæ], as distinguished from "down-feathers" [plumæ], are worked by muscles, sometimes too small to be easily detected, but amounting to four or five to each feather. In *Anas marila* and in *Sula bassana*, Nitzsch counted 3000 of the contour-feathers, so that in each bird there must have been 12000 of these muscles.

The down on newly hatched birds is only partial and temporary. The dry membranous bodies in the quill are the remains of the pulpy medulla of early life. Once or twice a year birds moult or renew their feathers; in many cases the difference between the summer and winter moults is very great.

In many groups the female is much less or very differently coloured to the male, the young birds are mostly coloured like the females; in the latter it is probably due to the arrested development of the colouring-matter, just as, on the other hand, old female birds sometimes partially assume the coloration of the male.

Peculiar formations of integument [corium] are known as caruncles, wattles, &c. The cere at the base of the bill is of the same nature.

All birds are oviparous; they have usually only one ovary, which lies towards the left side. The period of incubation varies from two weeks to nearly two months.

As to the nests of birds, Wallace draws the conclusion that " when both sexes are of strikingly gay and conspicuous colours, the nest is such as to conceal the sitting bird ; while, whenever there is a striking contrast of colours, the male being gay and conspicuous, the female dull and obscure, the nest is open and the sitting bird exposed to view." The Duke of Argyll, on the other hand, maintains that rather the nests themselves require concealment, and that the structure "which most completely covers up the eggs, or the sitting bird, may, and often does, render the nest itself only more conspicuous." It appears to be still a question whether birds build from instinct or from imitation. No bird hibernates.

Slight differences are sometimes observable between birds from different localities, differences in some cases considered to be specific ; but they are probably only " races " depending on local causes. According to Blanford, eastern and western " races " are found " passing into each other and breeding together where they meet in the Levant."

Systems of classification for birds have been numerous ; of the two latest—Huxley's and Garrod's—the first is founded on the palatal structure, the latter mainly on the disposition of the muscles of the thigh. Huxley divides birds into three orders— Saururæ, Ratitæ, and Carinatæ, the latter comprising all known living birds except the ostriches, emus, &c. The Carinatæ include four suborders:—(1) Dromæognathæ [vomer very broad, united in front with the maxillo-palatine plates, receiving behind the anterior ends of the pterygoid bones, &c.]: represented by *Tinamus* only ? (2) Schizognathæ [vomer various, tapers to a point arteriorly, behind it embraces the basisphenoidal rostrum between the palatines ; maxillo-palatines fissured] : includes plovers, gulls, cranes, Gallinæ, pigeons. (3) Desmognathæ [vomer abortive or very small ; "maxillo-palatines united across the middle line, either directly or by the intermediation of ossifications in the nasal septum "] : includes most of the Grallæ and Anseres which are not Schizognathous, the Accipitres, the Scansores, and many of the Volitores (*i. e.* swifts, kingfishers, hoopoes). (4) Ægithognathæ [vomer very broad, truncated in front and cleft behind, embracing the rostrum of the sphenoid between its forks]: includes the great majority of the Passerine birds. The last three " suborders " are divided into " groups."

Of this classification it has been remarked as " questionable " " how far any approach to a natural system can be based on the modification of one part of an animal's structure without any reference to other portions of it " (*Newton*).

Garrod divides birds into two subclasses according as the "ambiens" (or rectus femoris, a slender muscle arising above the acetabulum, and joining the tendon of the flexor perforatus digitorum) is present (Homalogonati) or absent (Anomalogonati). The former has four orders:—(1) Galliformes: including ostriches, Gallinæ, rails, cuckoos, parrots.   (2) Anseriformes: Anatidæ, penguins, petrels, &c.   (3) Ciconiiformes: storks, "Catharteæ," herons, Steganopoda, falcons, and owls.   (4) Charadriiformes: pigeons, plovers, cranes, gulls, &c.   The second subclass has three orders:—(1) Piciformes: woodpeckers, toucans, kingfishers, &c. (2) Passeriformes: Passeres, goatsuckers, rollers, &c. (3) Cypseliformes: swifts and humming-birds.

With regard to the palatal system a fifth suborder for the woodpeckers has been proposed by Parker [Saurognathæ]. He holds that "the lineaments of the old feathered fathers of the existent types" must be sought for in the embryo; and he finds that the parts of the Ægithognathous face, of which those of the Saurognathæ are but a "degradation and simplification," "were really built up of elements which had their true counterparts or 'symmorphs' in the snake."

Another system, in which external are preferred to internal or anatomical characters, was proposed by Sundevall in 1872.   The Passeres and Picæ of Linnæus form the "agmen" Psilopædes or Gymnopædes (young at first naked), comprising two orders, 19 cohorts, and 157 families. The second "agmen," Ptilopædes or Dasypædes (young covered with down from the first), comprises the orders Accipitres, Gallinæ, Grallatores, Natatores, Proceres, and Saururæ; excluding the last, these orders include 22 cohorts and 60 families.   In the Psilopædes each order is divided into two series, and in the cohorts of the first we have also 14 "Phalanges," each phalanx, like the cohorts, mostly bearing a character-name, such as Brevipennes, Latirostres, Novempennatæ, &c. The definitions are very clear, and all barbarous names of genera are rejected.   As this classification has been favourably received in some quarters, the following tabular view may be useful, and will enable it to be better understood.   Many of Sundevall's families are generally regarded as subfamilies.

Agmen primum, PSILOPÆDES.

Order 1. OSCINES.

Series prior, Lamelliplantares.

Cohorts.	*Phalanges.*	*Principal families.*
1. Cichlomorphæ.	1. Ocreatæ.	Lusciniinæ, Saxicolinæ, Turdinæ.

Cohorts.	*Phalanges.*	*Principal families.*
1. Cichlomorphæ.	2. Brevipennes.	Malurinæ, Copsychi- næ, Malaconotinæ, Troglodytinæ.
	3. Miminæ.	Vireoninæ, Phyllo- pneustinæ,Sylviinæ, Parinæ, Laniinæ.
	4. Ampelidinæ.	Oriolinæ, Campopha- ginæ, Dicrourinæ.
	5. Latirostres.	Ficedulinæ, Muscica- pinæ.
	6. Novempennatæ.	Motacillinæ, Icteri- inæ, Pardalotinæ.
2. Conirostres.	1. Ploceinæ.	Viduinæ,Accentorinæ.
	2. Amplipalatales.	Chloridinæ, Fringil- linæ.
	3. Atratipalatales.	Loxiinæ, Emberizinæ.
	4. Simplicirostres.	Ramphocelinæ, Tana- grinæ.
3. Coliomorphæ.	1. Novempennatæ.	Agelæinæ, Icterinæ.
	2. Humilinares.	Sturninæ, Bupha- ginæ, Fregilinæ.
	3. Altinares.	Nucifraginæ, Garru- linæ, Corvinæ.
	4. Idiodactylæ.	Subgarrulinæ (*Citta*), Paradisæinæ.
4. Certhiomorphæ.	....	Certhiinæ, Sittinæ.
5. Cynnirimorphæ.	....	Arbelorhininæ (*Cære- ba cyanea*), Drepa- nidinæ, Nectari- niidæ, Melipha- gidæ.
6. Chelidonimorphæ.	....	Hirundinidæ.

Series posterior, Scutelliplantares.

1. Holaspideæ.   Alaudinæ, Upupinæ.
2. Endaspideæ.   Furnariinæ, Dendrocolaptinæ.
3. Exaspideæ.   Oxyrhynchinæ, Tyranninæ, Todinæ, Piprinæ.
4. Pycnaspideæ.   Rupicolinæ, Ampelinæ.
5. Paietinæ.   Thamnophilinæ, Scytalopodinæ.

Order 2. VOLUCRES.

Series prior, Volucres zygodactyli.

Cohorts.	*Principal families.*
1. Psittaci.	Camptolophinæ (*Microglossus* &c.), Platycercini, Trichoglossini.
2. Pici angusticolles.	P. securirostres, P. ligonirostres, P. nudinares, Picumnini, Iynginæ.
3. Coccyges.	Ramphastinæ, Galbulinæ, Bucconinæ, Cuculinæ, Crotophaginæ.

Series posterior, Volucres anisodactyli.

4. Cœnomorphæ.	Musophaginæ, Coliinæ, Coraciinæ.
5. Ampligulares.	Trogoninæ, Podarginæ, Caprimulginæ, Cypselinæ.
6. Volucres longilingues.	(Humming-birds, divided into thirteen families.)
7. Volucres syndactylæ.	Meropinæ, Prionitinæ, Alcedininæ, Bucerotinæ.
8. Peristeroideæ.	Didinæ, Columbinæ, Megapeliinæ (*Goura* =*Lophyrus*).

Agmen secundum, PTILOPÆDES.

Order 3. ACCIPITRES.

1. Nyctharpages.	Ululinæ, Buboninæ, Noctuinæ.
2. Hemeroharpages.	Asturinæ, Buteoninæ, Falconinæ, Aquilinæ, Milvinæ.
3. Saproharpages.	Gypætinæ, Vulturinæ.
4. Necroharpages.	Cathartinæ, Polyborinæ, Dicholophinæ.

Order 4. GALLINÆ.

1. Tetraonomorphæ.	Pteroclinæ, Tetraoninæ.
2. Phasianomorphæ.	Phasianinæ, Pavoninæ, Perdicinæ, Hemipodiinæ.
3. Macronyches.	Catheturinæ (*Catheturus*=*Talegallus*), Megapodiinæ.
4. Duodecimpennatæ.	Cracinæ, Penelopinæ.
5. Struthioniformes.	Crypturinæ.
6. Subgrallatores.	Thinocorinæ, Chionodinæ.

Order 5. GRALLATORES.

Series prior, Grallatores altinares.

Cohorts.	Principal families.
1. Herodii.	Ardeinæ.
2. Pelargi.	Plataleinæ, Ciconiinæ, Ibidinæ, Scopinæ.

Series posterior, Grallatores humilinares.

3. Limicolæ.	Totaninæ.
4. Charadriinæ.	Charadriinæ, Otidinæ, Gruinæ, Rallinæ.

Order 6. NATATORES.

1. Longipennes.	Sterninæ, Larinæ.
2. Pygopodes.	Alcariæ, Colymbinæ.
3. Totipalmatæ.	Pelecaninæ.
4. Tubinares.	Procellariinæ, Diomedeinæ.
5. Impennes.	Spheniscinæ.
6. Lamellirostres.	Phenicopterinæ, Anatinæ.

Order 7. PROCERES.

1. Proceres veri.	Struthioninæ, Dromæinæ.
2. Subnobiles.	Apteryginæ.

Order 8. SAURURÆ.

(Archæopteryx.)

In the more simple arrangement of Schmarda (1878) there are ten orders :—1. Archæopterygida ; 2. Natatores ; 3. Grallatores ; 4. Cursores ; 5. Gallinacea ; 6. Columbæ ; 7. Passeres ; 8. Syndactyli ; 9. Scansores ; and 10. Raptatores. This arrangement differs from that of Claus (1876) in that Syndactyli are combined with Passeres, and the Cursores are placed last after Raptatores.

In the absence of an absolute agreement among ornithologists, the classification here adopted so far differs from the ordinary modifications of the "mercurial taxonomist" as to include the " Volitores" of Owen as an order, and in beginning with the Pici as the lowest type of living birds and ending with Psittaci as the highest. The "Reptilian birds" are aberrant, and, not forming a direct passage from the preceding class, are placed last.

The characters of birds are so unvarying " that it is difficult to separate them into subordinate groups ;" these are " more arbitrary and artificial than in those of the other vertebrate classes."

Bill without teeth (all living birds).
Breast-bone keeled.
Feet not webbed.
Legs feathered to the knees.
Nostrils pierced in the bony structure of the bill.
Bill various, not arched from the base.
Tongue long, extensible .... PICI.
Tongue short, not extensible.
One toe behind.
Feet short, weak .... VOLITORES.
Feet long, moderately strong... PASSERES.
Two toes behind .... SCANSORES.
Bill strongly arched from the base.
Two toes behind .... PSITTACI.
One toe behind .... ACCIPITRES.
Nostrils pierced in a membrane.
Hind toe on a level with the others ... COLUMBÆ.
Hind toe elevated .... GALLINÆ.
Legs naked above the knees .... GRALLÆ.
Feet webbed... ANSERES.
Breast-bone not keeled .... STRUTHIONES.
Bill with teeth. Extinct. (*Odontornithes.*)
Breast-bone keeled .... ODONTORMÆ.
Breast-bone not keeled .... ODONTOLCÆ.
Breast-bone rudimentary .... SAURURÆ.

There are about 8000 species of birds according to Schmarda (1878), but 11,162 are enumerated in G. R. Gray's ' Hand-List ' (1871); of these, about 270 species, not counting stragglers, are British.

Order I. PICI.

SAGITTILINGUES. CELEOMORPHÆ. SAUROGNATHÆ.

Bill straight, wedge-shaped : tongue extensible, barbed at the end. Tail-feathers stiff at the points. Feet short, stout; first and fourth toes turned backwards.

The bill in the typical species is of an ivory-like hardness and much compressed, especially towards the tip. The tongue, provided with a tenacious secretion, is capable of great elongation and extension, the two cornua of the hyoid bone extending round to the back of the head, forming a bow which can be lengthened or shortened by accompanying muscles.

The palatal structure is "at a most simple and Lacertian stage" (*Parker*). The vomers are delicate rod-like bones which, in some cases, remain permanently separate. The quadrate bone is very short. The sternum has two notches on each side posteriorly, and a forked manubrial process. There is only one carotid. The tarsi are covered with short imbricated scales anteriorly. In *Sasia* and *Apternus* the inner posterior toe is wanting.

The Pici are shy untamable birds, flying from man, living in woods, and nesting in holes of trees; the males share the duty of incubation. They live on insects and worms, or on fruit; some of the North-American species of *Melanerpes* are said to enter dovecots for the purpose of sucking the eggs of pigeons. They creep, rather than climb, on the trunks and branches of trees. Picumnidæ are exceptional; their tail-feathers are broad and rounded at the tip, and they do not appear to climb.

Sundevall was the first (but only for a time) to separate these birds as an order, in which he is followed by Carus and Huxley. To it belongs the green woodpecker (*Gecinus viridis*), great spotted woodpecker (*Picus major*), black woodpecker (*Dryocopus martius*), and the wryneck (*Yunx torquilla*).

There are 350 species in G. R. Gray's 'List;' but none are found in Madagascar or in Australia.

Picumnidæ.	Chrysoptilus.	Hemilophus.
Picumnus.	Chrysonotus=Tiga.	Dryocopus.
Sasia.	Gecinus.	Campophilus.
	Celeus.	Picus.
Picidæ (Wood-	Meiglyptes.	
peckers).	Colaptes.	*Yungidæ.*
Apternus=Picoides.	Geocolaptes.	Yunx (Wryneck).
	Melanerpes.	

Order II. VOLITORES.

FISSIROSTRES.   PICARIÆ.   STRISORES.   COCCYGOMORPHÆ.

Bill various, with mostly a wide gape; no cere. Legs small and weak; a back toe, outer toe sometimes reversible. Wings strong.

In this order the bill is often remarkable for length or breadth, or both. The feet are mostly small, with little grasping power.

The wings are strong, and frequently long and pointed, these birds "moving solely by flight" and taking their food on the wing. They nestle in holes of trees, or in holes of banks, or on the earth; or, as some of the swallows, they build mud nests against the sides of rocks or walls. Humming-birds construct very delicate and compact nests of moss, lichens, feathers, &c.; while the kingfisher is content to heap together a few fish-bones, in some hole or hollow, on which to lay her eggs. In the hornbills the male, as is well known, shuts up the female in her nest in a hollow tree by filling up the entrance with mud, leaving just enough space through which the latter and her family may receive the food he brings them. Livingstone says that the "poor slave of a husband" often dies of inanition after his exertions.

The majority of these birds live on insects; humming-birds in part on the nectar in flowers, which they extract without alighting; kingfishers feed on water-insects, small fishes and their fry; hornbills on fruits, and sometimes on small birds.

Among other well-known birds in this order we have the kingfisher (*Alcedo ispida*), the laughing jackass of Australia (*Dacelo gigantea*), the nightjar or goatsucker (*Caprimulgus europæus*), whip-poor-will of North America (*Antrostomus vociferus*), the "old jew" of New South Wales (*Podargus auritus*), sand-martin (*Cotyle riparia*), house-martin (*Chelidon urbica*), swallow (*Hirundo rustica*), the "edible" swallow (*Collocalia esculenta*, whose nests, partially composed of a gelatinous secretion of the bird itself, is much sought after by Chinese gourmands), and the swift (*Cypselus apus*). Bee-eaters and rollers also belong to this order. The humming-birds (of which there are 460 species) Sundevall, under his cohort "Volucres longilingues," splits up into twelve families. Their so-called genera are about 150.

Volitores include the Coccygomorphæ of Huxley except the zygodactyle families and Coliidæ, and the whole of his Cypselomorphæ (Macrochires, *Nitzsch*). The Fissirostres of Cuvier were confined to the genera *Hirundo, Cypselus, Caprimulgus*, and *Podargus*. Syndactyli of the same author comprised the five genera *Merops, Prionites, Alcedo, Ceyx*, and *Todus*. Picariæ are an indefinite group not generally adopted. Hirundinidæ are occasionally placed in the Passeres; Wallace says that they "are undoubtedly very isolated;" but Huxley thinks that they are "very nearly related" to the Cypselidæ. Claus and Schmarda unite them and Caprimulgidæ in one "group" (Fissirostres). The order is included in the "Volucres anisodactyli" of Sundevall; but he includes in it also the Pigeons.

Capitonidæ (Barbets).

Pogonorhynchus=
    Laimodon.
Megalæma.
Capito.

Coraciidæ.

Eurylæmus.
Corydon.
Eurystomus.
Coracias (Roller).
Colaris.

Bucerotidæ (Horn-
    bills).

Bucorax.
Buceros.
Euryceros.

Upupidæ.

Upupa (Hoopoe).
Irrisor=Promerops.

Meropidæ.

Nyctiornis.
Merops (Bee-eater).

Galbulidæ.

Jacamerops=Lam-
    protila.
Galbula.

Alcedinidæ.

Ceyx.
Syma.
Halcyon.
Tanysiptera.
Alcedo (Kingfisher).
Dacelo.

Todidæ.

Todus.

Prionitidæ.

Prionites=Momotus.

Trogonidæ.

Hapaloderma.
Priotelus.
Harpactes.
Calurus.
Trogon.

Caprimulgidæ.

Antrostomus.
Steatornis.
Chordeiles.
Ægotheles.
Caprimulgus (Goat-
    sucker).
Podargus.
Nyctibius.
Batrachostomus.

Hirundinidæ.

Progne.
Cotyle(Sand-Martin).
Chelidon (House-
    Martin).
Hirundo (Swallow).
Atticora.

Cypselidæ.

Chætura.
Collocalia.
Cypselus (Swift).

Trochilidæ (Hum-
    ming-birds).

Ramphodon.
Phaëthornis.
Campylopterus.
Petasophora.
Thalurania.
Sparganura=
    Cometes.
Heliothrix.
Lophornis.
Glaucis.
Selasphorus.
Trochilus.
Amazilis.
Docimastes.
Patagona.

Order III. SCANSORES.

ZYGODACTYLI. AMPHIBOLI. ERUCIVORES. COCCYGES. PICARIÆ. COCCYGOMORPHÆ.

Bill various, but never arched from the base; no cere. Tongue not extensile. Tarsi with broad scutes. First and fourth toes turned backwards.

The bill varies more in size than in shape, being very small comparatively in the cuckoos, and nearly as large as the bird itself in some of the toucans. With the latter it is of a delicately cellular structure, and therefore very light.

The Scansores do not climb, or rather creep, in the sense of woodpeckers and creepers; some of them live on the ground and are good runners; they have a short flight, and place their nests in holes of decaying trees, or, as in some of the cuckoos, they lay their eggs in the nests of other birds. *Indicator*, however, builds a complex bottle-shaped nest. They feed mostly on insects and fruits, the toucans and some of the ground-cuckoos also on small birds and reptiles: the former are known to "regurgitate partially digested food, and after submitting it to a rude kind of mastication by their enormous beaks, again to swallow it."

The cuckoo (*Cuculus canorus*) is the only British bird belonging to this order, which includes also the plantain-eaters and the toucans. Coliidæ are South-African birds with no obvious allies; in them all the toes are turned forwards. Murie thinks that they are an annectent group between "Coccygomorphæ and Coracomorphæ," which he names Coliomorphæ (not the Coliomorphæ of Sundevall). G. R. Gray placed them in the Conirostres. They have the peculiar habit of hanging by one foot with the head downwards.

This order corresponds to the second group of Huxley's Coccygomorphæ, except that he includes Galbulidæ. To this second group he thinks it may be desirable to restrict the term. The order also forms part of the Zygodactyli and of the Picariæ of the older authors, which included the Volitores and woodpeckers, and for some writers the parrots and chatterers also.

Musophagidæ.
Corythaix (Touraco).
Schizorhis.
Musophaga (Plantain-eater).

Cuculidæ.
Scythrops.
Eudynamys.
Cuculus (Cuckoo).
Lamprotornis = Chrysococcyx.

Leptosomus.
Phœnicophaes.
Crotophaga (Ani).
Corydonyx (Sericosomus).
Coccyzus.
Saurothera.
Centropus.
Polophilus.

———

Indicator (Honeyguide).

———

Bucconidæ (Puffbirds).
Brachypetes = Chelidoptera.
Monasa.
Malacoptila.
Bucco.

Rhamphastidæ (Toucans).
Pteroglossus.
Rhamphastos.

Coliidæ.
Colius.

Order IV. PASSERES.

Bill various, but never arched from the base; no cere. Tongue not fleshy. Legs moderate; one toe behind, three in front, the outer joined at its base to the middle toe.

The comparatively strong feet are formed for perching, all the toes being on the same level, and only one directed backwards. The tarsus is covered anteriorly by five or six or seven imbricated scales (scutes or scutellæ), rarely by one only. The females are smaller than the males, and are less distinctly coloured. They build complex nests, and the young leave the egg in a blind and naked state.

Passerine birds, as here limited, have only moderate powers of flight, and live chiefly either on berries and seed, or on insects and worms; but a few are omnivorous. This order contains all the singing birds; they have all, even those species that do not sing, a lower larynx, which may be of "every degree of complexity" and worked by five pairs of muscles, or it may only have two pairs, or may be devoid of any. Wallace observes that Passeres with "imperfect singing apparatus" are characterized by having wings with ten primaries, while all other birds have nine only, or if ten then the first "below its proportionate size."

Besides the names mentioned below, the following are some of the most familiar birds belonging to this order :—the nightingale (*Luscinia philomela*), [the Persian nightingale or bulbul is the *Luscinia Hafizii*, the bulbuls of India are *Pycnonoti*], blackcap(*Curruca atricapilla*), white-throat (*Curruca cinerea*), garden-warbler (*Sylvia hortensis*), wood-wren (*Phyllopneuste sibilatrix*), chiff-chaff (*Phylloscopus rufus*), robin (*Erythacus rubecula*), hedge-sparrow (*Accentor modularis*), titlark (*Anthus pratensis*), wheatear (*Saxicola œnanthe*), thrush (*Turdus musicus*), blackbird (*Turdus merula*), fieldfare (*Turdus pilaris*), wagtail (*Motacilla alba*), wren (*Troglodytes vulgaris*), goldfinch (*Carduelis elegans*), canary (*Carduelis canaria*), chaffinch (*Fringilla cœlebs*), hawfinch (*Coccothraustes vulgaris*), sparrow (*Passer domesticus*), linnet (*Linota cannabina*), yellow-ammer (*Emberiza citrinella*), ortolan (*Emberiza hortulana*), skylark (*Alauda arvensis*), crow (*Corvus corone*), raven (*Corvus corax*), jackdaw (*Corvus monedula*), rook (*Corvus frugilegus*), magpie (*Pica caudata*), and jay (*Garrulus glandarius*).

The lyre-bird (*Menura superba*) is a very aberrant form, and has been placed with the Gallinæ. Its affinities are supposed to be with Pteroptochidæ or with *Orthonyx* (a gigantic wren in other words), and also with the Birds of Paradise. *Todus* is another peculiar form ; its nearest living allies, according to Murie, after an examination of its *skeleton*, are the motmots and kingfishers. Claus, however, confines it to its old place among the Tyrannidæ.

Although now much restricted (opinions differing as to its extent), the Passeres are still the most numerous of all the orders of birds, and, owing to their slightly varying characters, one of the most difficult to classify. Wallace, recognizing four typical forms of wings, proposes to classify them thus :—primaries 10, the first well developed (4), or "reduced" (1), or rudimentary (3), and primaries 9 only (2). The typical or "Turdoid series" (1) contains the great majority of families, the "Tanagroid series" (2), the "Sturnoid series" (3), and "Formicarioid series" (4) comprise the remainder. The Passeres have also been divided into "Acromyodi," in which the intrinsic muscles of the voice-organs are fixed to the end of the bronchial half-rings, and "Mesomyodi," in which they join them at or near the middle.

Taking the form of the bill, as is most commonly done, as a practical character correlated in many respects with what is best known to the ordinary observer, and omitting Fissirostres and Levirostres, which are included in the Volitores, we have four suborders :—

Bill elongate, slender, often curved.................. TENUIROSTRES.
Bill shorter and stouter, never curved.
   Bill notched at the tip............................ DENTIROSTRES.
   Bill entire, or only obsoletely notched.
      Bill short, conical .... CONIROSTRES.
      Bill large, subconical or compressed .... MAGNIROSTRES.

TENUIROSTRES.—Bill lengthened, slender, mostly curved, never notched at the tip. Legs strong.

Perching or "climbing" birds, living on small insects, larvæ, &c., which they do not catch on the wing. They have no song.

Nectariniidæ.	Zosterops.	Meliphaga.
Dicæum.	Promerops.	Anthornis.
Anthreptes.		Anthochæra.
Arachnothera.	*Meliphagidæ.*	Prosthemadera
Nectarinia = Cinnyris	Myzomela.	(Parson-bird).
(Sun-bird).	Glyciphila.	Tropidorhynchus.

Dendrocolaptidæ.
Anabates.
Xenops.
Lochmias.
Oxyrhynchus.
Synallaxis.
Sittasomus.
Dendrocolaptes.
Xiphorhynchus.

Geositta.
Furnarius = Opetio-
    rhynchus.

Certhiidæ.
Tichodroma.
Climacteris.
Certhia (Creeper).

Sittidæ.
Sitta (Nuthatch).
Orthonyx.

Cærebidæ.
Diglossa.
Dacnis.
Cæreba.
Drepanis.

DENTIROSTRES.—Bill slenderly conical, the upper mandible notched or toothed, sometimes hooked at the tip.

Mostly worm- or insect-feeders, a few preying on birds and small mammals, or on berries, the insects frequently taken on the wing.

Our best-known songsters—nightingale, thrush, blackcap, &c. —belong to this group.

Pteroptochidæ.
Rhinocrypta.
Hylactes.
Scytalopus.
Pteroptochus.

Formicariidæ (Ant-
    birds).
Thamnophilus.
Formicarius.
Pithys.
Conopophaga.
Grallaria.

Pittidæ.
Pitta.

Timaliidæ.
Cissa = Kitta.
Eupetes.
Cinclosoma.
Timalia.
Paradoxornis.

Tyrannidæ.
Megarhynchus.

Saurophaga.
Myiarchus.
Gubernetes.
Tyrannus.
Fluvicola.
Alectrurus.

Dicruridæ.
Irena.
Artamus.
Dicrurus = Edolius
    (King-crow).

Laniidæ.
Falcunculus.
Pachycephala.
Rectes.
Eopsaltria.
Myiolestes.
Malaconotus.
Lanius (Shrike).
Vireo.

Campephagidæ.
Pericrocotus.

Campephaga =
    Ceblepyris.

Muscicapidæ.
Monarcha.
Myiagra.
Petrœca.
Rhipidura.
Muscicapa (Fly-
    catcher).

Mniotiltidæ.
Mniotilta.
Icteria.
Dendrœca.
Setophaga.

Sylviidæ.
Accentor (Hedge-
    Sparrow).
Acanthiza.
Regulus (Gold-crest).
Curruca (White-
    throat).
Luscinia (Night-
    ingale).

Sylvia.
Phyllopneuste.
Phylloscopus.
Saxicola (Stonechat, &c.).
Sialis.
Copsychus.
Dasyornis.
Drymœca.
Cisticola.
Malurus.

Turdidæ.

Ixos.

Andropadus.
Pycnonotus.
Pomatorhinus.
Mimus.
Turdus (Thrush, &c.).
Cinclus (Dipper).

Motacillidæ.

Seiurus.
Budytes.
Motacilla (Wagtail).
Henicurus.
Ephthianura.

Grallina.
Anthus (Tit-lark).

Troglodytidæ.

Troglodytes (Wren).
Thryothorus.

Paridæ.

Panurus=Calamophilus.
Acredula=Mecistura.
Parus (Tit).

CONIROSTRES.—Bill short, stoutly conical, not, or occasionally slightly, notched, and never hooked at the point.

Mostly granivorous, and more exclusively perchers. They are all of comparatively small size, and many are eminently song-birds.

Crows, starlings, and even hornbills have been referred to this group. Phytotomidæ is sometimes placed in the next group. The Eocene *Protornis* is one of the earliest known Passerine birds.

Tanagridæ.

Euphonia.
Calliste.
Pyranga.
Tanagra.
Arremon.
Tachyphonus.
Nemosia.
Saltator.
Pitylus.
Bethylus.

Ploceidæ.

Symplectes (Weaver-bird).
Ploceus.

Amadina.
Estrelda.
Pyrenestes.
Oryzornis (Paddy-bird).
Vidua (Whydah-bird).

Fringillidæ.

Cardinalis.
Pyrrhula (Bullfinch).
Loxia (Crossbill).
Corythus (Hawfinch).
Coccothraustes (Greenfinch).
Passer (Sparrow).

Fringilla (Finch).
Carduelis (Gold-finch).
Linota (Linnet).
Zonotrichia.
Emberiza (Bunting).

Phytotomidæ.

Phytotoma.

Alaudidæ (Larks).

Alauda.
Otocorys.

———

*Protornis.

MAGNIROSTRES.—Bill large, elongate, conical, not or only slightly notched.

Q

Frequently ground-feeders and often omnivorous. They have no song, but some are mimics. The raven is held by Swainson to be the most typical of birds. *Bombycilla* is placed by some in the Muscicapidæ. There is considerable difference of opinion as to the limits of the families.

This group agrees with the Magnirostra of Schmarda, and includes the major part of the Coliomorphæ of Sundevall.

Ampelidæ (Chatterers).

Ampelis=Cotinga.
Phibalura.
Procnias.
Chasmorhynchus.
Gymnocephalus.
Coracina.
Calyptomena.
Rupicola (Cock-of-the-rock).
Pipra (Manakin).
Calyptura.
Pardalotus.
Tityra=Psaris.

Bombycillidæ.

Bombycilla (Waxwing).

Oriolidæ.

Chlamydodera.
Sericulus.
Ptilonorhynchus.
Oriolus (Oriole).

Sturnidæ.

Dolichonyx.
Icterus.
Cassicus.
Molothrus (Cowbird).
Quiscalus.
Gracula.
Gymnops.
Acridotheres.
Buphaga.
Heteralocha=Neomorpha.
Pastor.
Lamprotornis.
Sturnus (Starling).

———

Fregilupus.

Corvidæ.

Struthidea.
Glaucopis.
Scissirostrum.
Cracticus=Barita.
Garrulus (Jay).

Lophocitta.
Gymnorhina.
Podoces.
Pyrrhocorax.
Fregilus (Chough).
Nucifraga (Nutcracker).
Corvus (Crow, &c.).
Pica (Magpie).

Paradiseidæ (Birds of Paradise).

Epimachus.
Ptilorhis.
Seleucides.
Paradisæa.
Cicinnurus.
Astrapia.
Lophorina.
Semioptera.
Parotia.
Diphyllodes.

———

Menura.

Order V. COLUMBÆ.

GEMITORES. GYRANTES. BIPOSITORES. PERISTEROMORPHÆ.

Bill various; nostrils pierced in a membrane. Hind toe placed on a level with the other toes; claws short, only slightly curved. Wings strong.

The bill is "swollen at the tip, and provided at the base with a tumid membranous space, in which the nostrils open;" this part often assumes a warty appearance. The feet are as much fitted for walking as perching; but the tarsus is short and stout, scutellate in front, and sometimes feathered.

Pigeons, the best known family of this order, are monogamous, and pair for life. They are without a gall-bladder, except in *Carpophaga* and *Ptilopus*, and hence, it is said, their placid disposition; it is wanting, however, in many other birds. Their nests are very simple, a few sticks, or merely a hollow amongst herbage. The males and females sit by turns on the eggs. The two dodos, *Didus ineptus* and *Didus solitarius*, are now extinct. They were very heavy birds, incapable of flight, and in other respects very unlike pigeons. *Didunculus* is an intermediate form.

Among the members of this order are the common pigeon, supposed to be descended from the rock-dove (*Columba livia*), wood-pigeon (*Columba palumbus*), stock-dove (*Columba œnas*), turtle-dove (*Turtur auritus*), passenger-pigeon (*Ectopistes migratorius*), the bronze-wing (*Phaps chalcoptera*), and the great-crested pigeon of the Indian Archipelago (*Goura coronata*).

This order with Megapodiidæ and Cracidæ were the Pullastreæ of Sundevall, but in 1872 it became his eighth cohort (Peristeroideæ) of the Volucres. Didunculidæ and Dididæ are united as a family—Didina. *Goura* is the type of another family—Megapeliinæ. *Calœnas* has also been separated as a family. Pteroclidæ are by some writers considered to belong to this order.

Columbidæ (Pigeons).	Phaps.	Zenaida.
Calœnas.	Geopelia.	Ptilopus.
Goura = Lophyrus = Megapelia.	Turtur (Turtle-dove).	*Didunculidæ.*
Treron.	Œna.	Didunculus.
Vinago.	Ectopistes.	
Carpophaga.	Chamæpelia.	*Dididæ.*
Columba.	Macropygia.	*Didus (Dodo).

Order VI. GALLINÆ.

RASORES. CLAMATORES. ALECTOROMORPHÆ.

Bill mostly rather short, vaulted, edges of the upper mandible overlapping. Nostrils pierced in a membranous space. Hind toe above the level of the others; claws thick, obtuse.

The legs are very stout, feathered to the knees and often beyond them; the tarso-metatarsus at its back part is often armed with a spur, or accessory toe, rarely with two. The head is small, the body bulky, the wings only adapted for short flight, which is accompanied by a whirring sound.

In the Tinamous the sternal notch extends nearly to the costal margin.

The Gallinæ are mostly polygamous; the females make very slight nests on the ground; the young feed themselves. Megapodiidæ raise huge mounds in which they place their eggs, leaving them to be hatched by the heat of the sun.

The order Turnicomorphæ of Huxley is confined to the family Turnicidæ.

Tinamus represents the Dromæognathous suborder, which Huxley considers to be nearly allied to the Struthiones. *Opisthocomus* comprises his Heteromorphæ. *Chionis* has been placed with the pigeons; some consider it to be allied to the rails, and others to the grouse; it has also been suggested that from such a stock gulls and plovers have descended. Carus includes *Thinocorus* in the Chionidæ.

This order contains the quail (*Coturnix dactylisonans*), partridge (*Perdix cinerea*), ptarmigan (*Lagopus vulgaris*), red grouse (*Lagopus scoticus*), black grouse (*Tetrao tetrix*), prairie-hen of America (*Tetrao cupido*), capercailie (*Tetraogallus urophasianus*), guinea-fowl (*Numida meleagris*), peacock (*Pavo cristatus*), and turkey (*Meleagris gallo-pavo*) [the "turkey" of Australia is *Otis australasianus*]. The domestic fowl is supposed to be descended from the jungle-cock (*Gallus bankiva*).

Pteroclidæ.
Syrrhaptes.
Pterocles.
———
Thinocorus.

Tetraonidæ.
Cryptonyx.
Tetraogallus.
Ortyx.
Odontophorus.
Coturnix (Quail).
Perdix (Partridge).
Lagopus.
Tetrao (Grouse).

Turnicidæ.
Turnix=Hemipodius.
Pedionomus.

Phasianidæ.
Lophophorus.
Euplocamus.
Gallus (Fowl).
Tragopan.
Numida (Guinea-fowl).
Meleagris (Turkey).
Pavo (Peacock).

Polyplectron.
Argus.

Megapodiidæ.
Leipoa.
Talegalla.
Megapodius.

Cracidæ.
Urax=Mitu=Pauxi.
Crax (Curassow).
Ortalida.
Penelope (Guan).
Oreophasis.

Opisthocomidæ.	*Tinamidæ.*	Rhynchotus.
Opisthocomus.	Nothura.	Tinamotis.
	Tinamus = Crypturus	
	(Tinamou).	

Chionidæ.

Chionis.

Order VII. STRUTHIONES.

RATITÆ. BREVIPENNES. PLATYSTERNÆ. PROCERI.

Bill various. Breast-bone without the mesial keel. Wings rudimentary. Barbs of the feathers not connected to one another.

All other birds have a keeled breast-bone; and in no others are the feathers so hair-like; they have no accessory plumes in the ostrich and *Apteryx*, while in the cassowary there are two. The quill-feathers of the wings and tail are the ostrich-feathers of commerce. Few air-cells are found in the bones. There is no lower larynx. The feet are large; the toes are two or three, there being no back toe, except a rudimentary one in *Apteryx*. In the ostrich alone are the two pubic bones united, and it has no, or only rudimentary, clavicles. In the fœtal state of this bird there are 18 caudal vertebræ, in the adult they are reduced to 9. In the emu, in both sexes, there is a large membranous sac opening into the trachea; its use is unknown. *Apteryx* has its nostrils placed at the tip of the bill.

The intestinal canal is very complicated in the ostrich; the cæca are upwards of two feet in length, but they are absent in the cassowaries. The stomach or gizzard is very muscular.

Ostriches are polygamous; both parents attend to their eggs, which are laid in holes in the sand. The nandou (*Rhea americana*) is also polygamous, takes the water readily, and is a good swimmer. Emus and cassowaries are monogamous. *Apteryx* is nocturnal. *Dinornis* and *Æpyornis*, both extinct, are amongst the largest of known birds. The egg of *Æpyornis maximus* was equal to about 148 hen's eggs, while the egg of the apteryx is one quarter the weight of the bird.

The following short list includes all the well-ascertained living forms of what probably has been a very extensive group :—The ostrich (*Struthio camelus*), three nandous (*Rhea americana, R. darwinii,* and *R. macrorhyncha*), two emus (*Dromæus novæ-hollandiæ* and *D. irroratus*), five cassowaries (*Casuarius galeatus,*

C. australis, C. bennettii [the mooruk], *C. uniappendiculatus*, and *C. bicarunculatus*), and three kiwis (*Apteryx australis, A. owenii.* and *A. mantelli*). Six other species, or so-called species, of cassowaries have been described, but they appear to differ very slightly from one or another of the above (in one case the young), and are mostly only known from one or two specimens.

Struthiones have been combined with bustards, and even with the dodo, to form the order Cursores of the older authors. They form, according to Huxley, one of the two orders of living birds. Claus places them after Accipitres, as the last order of birds. *Æpyornis* has been referred to the Vulturidæ.

Struthionidæ.	*Casuariidæ.*	*Æpyornithidæ.*
Struthio (Ostrich).	Dromæus (Emu).	*Æpyornis.
Rheidæ.	Casuarius (Cassowary).	*Apterygidæ.*
Rhea (Nandou).		Apteryx (Kiwi).
	Dinornithidæ.	
	*Dinornis (Moa).	

Order VIII. GRALLÆ.

GRALLATORES.   LITTORES.

Bill various, generally longer than the head. Legs long, naked above the knee; toes not webbed.

The tibia and tarso-metatarsal bone are generally long, and the former more or less naked. The neck and bill are also long; the latter may be weak or porous, as in the snipes, or strong, with a cutting-edge, as in the herons. The body is mostly thin and compressed, with wings of great power, although the flight is often slow, and near the carpal joint they are sometimes armed with a spur or spine. Many can swim with ease; but the greater part are waders, living on fish, worms, &c; a few feed on grain, insects, &c., never going near the water.

In some of the Grallæ the trachea is singularly convoluted, bent sharply back in the thorax before it enters the lungs, or, as in the crane, the convoluted part is lodged between the two walls of the keel of the breast-bone.

These birds construct very simple nests, some indeed, as the bustards, lay their eggs on the bare ground. The young in some cases feed themselves, especially when the parents are polygamous, In many species there is a considerable difference between summer and winter plumage.

In this order we find the crane (*Grus cinerea*), coot (*Fulica atra*), corn-crake (*Crex pratensis*), peewit, lapwing, or plover of the London poulterers (*Vanellus cristatus*), dotterel (*Charadrius morinellus*), golden plover (*Charadrius pluvialis*), curlew (*Numenius arquatus*), whimbrel (*Numenius phæopus*), snipe (*Scolopax gallinago*), jack-snipe (*Scolopax gallinula*), woodcock (*Scolopax rusticola*), knot (*Tringa canutus*), sacred ibis (*Ibis religiosa*), red ibis (*Tantalus ruber*), jabiru (*Mycteria americana*), adjutant (*Mycteria argala*), stork (*Ciconia alba*), bittern (*Botaurus stellaris*), and heron (*Ardea cinerea*). Sandpipers are various species of *Totanus* and *Tringa*.

The last four families in the following list are sometimes separated as a distinct order [Ciconiæ=Pelargomorphæ]. *Palamedea*, "a lacertine goose," according to Parker, has been placed among the Natatores.

Otididæ (Bustards).
Otis.
Eupodotis.

Psophiidæ.
Psophia.

Gruidæ (Cranes).
Anthropoides.
Balearica.
Grus.

Parridæ.
Parra (Jacana).

Rallidæ.
Ocydromus.
Notornis.
Porphyrio.
Podoa=Heliornis.
Fulica (Coot).
Gallinula (Water-hen).
Crex (Corn-crake).
Rallus (Rail).

———

Mesites.

Rhinochetidæ.
Eurypyga.
Rhinochetus.

Charadriidæ.
Strepsilas (Turnstone).
Vanellus (Plover, Lapwing).
Ægialites.
Charadrius (Plover, Dotterel).
Cursorius.
Œdicnemus.
Esacus.
Hæmatopus (Oyster-catcher).

———

Glareola (Pratincole).

Scolopacidæ.
Calidris (Sanderling).
Tringa (Sandpiper).
Totanus (Redshank, &c.).

Machetes (Ruff ♂, Reeve ♀).
Phalaropus.
Limosa (Godwit).
Himantopus.
Numenius (Curlew, &c.).
Recurvirostra (Avocet).
Scolopax (Snipe, &c.).
Rhynchæa.
Dromas.

Tantalidæ.
Tantalus.
Geronticus.
Falcinellus.
Ibis.
Platalea (Spoonbill).

Ciconiidæ.
Mycteria.
Anastomus.
Ciconia (Stork).

Scopidæ.
Scopus.
Balæniceps.

Ardeidæ.	Nycticorax.	Tigrisoma.
Cancroma (Boatbill).	Botaurus (Bittern).	Ardea (Heron, Egret).

Palamedeidæ.

Opistholophus =
　Chauna.
Palamedea.

Order IX. ANSERES.

NATATORES.

Legs mostly short, naked above the knee; toes webbed, the hind, toe, except in the Steganopoda, free.

The legs are placed behind the centre of gravity, so that many can only walk in an erect position. The body is heavy, and covered with a thickish coating of down beneath the feathers. The anterior toes only are webbed, except in the Steganopoda, which have the hind toe also united to the others; in *Podiceps* the web is confined to the sides of the toes. The bill is variously formed, flat, and furnished with lamellæ in the geese, with tubular nostrils in the petrels, compressed in the gulls, and provided with a gular pouch in the pelicans. In some of the males of the duck and merganser families there is a bony dilatation of the lower part of the trachea (two in the sheldrake).

The Anseres are almost invariably water-birds; they are mostly polygamous; the young are often capable of feeding themselves. The nest, if any, is always of the rudest kind, and they lay few eggs. The penguins are incapable of flight, but, aided by their fin-like wings, they are the best of divers and swimmers; on the other hand, the albatross seems to pass its whole life in the air, now motionless, now "performing its vigorous evolutions without a perceptible movement of the wings," feats "simply impossible by any mechanical means of which we have the least conception" (*Wyville Thomson*). Besides fish, many feed on seaweed, grass, &c.

This order contains the wild swan or hooper (*Cygnus ferus*), the tame swan (*Cygnus olor*), wild goose (*Anser ferus*), wild duck (*Anas boschas*), widgeon (*Anas penelope*), teal (*Querquedula crecca*), eider (*Somateria mollissima*), sheldrake (*Tadorna vulpanser*), pochard, or dun-bird (*Fuligula ferina*), scaup (*Fuligula marila*), golden-eye (*Clangula glaucion*), gannet (*Sula bassana*), booby (*Sula fusca*), shag (*Phalacrocorax graculus*), cormorant (*Phalacrocorax carbo*), pelican (*Pelecanus onocrotalus*), frigate-bird

(*Tachypetes aquila*), tropic bird (*Phaëthon phœnicurus*), stormy petrel (*Thalassidroma pelagica*), shearwater (*Puffinus cinereus*), fulmar (*Procellaria glacialis*), albatross (*Diomedea exulans*), grey gull (*Larus canus*), noddy (*Anous stolidus*), grebe (*Podiceps cristatus*), dabchick (*Podiceps minor*), puffin, coulterneb, or pope (*Mormon fratercula*), king-penguin (*Aptenodytes patagonica*), and jackass-penguin (*Eudyptes demersus*). The great auk (*Alca impennis*) is extinct.

It is usual to divide the Anseres into four groups, which some authorities rank as orders; these are:—Lamellirostres (=Unguirostres—Chenomorphæ): edges of the bill furnished with thin plates or lamellæ. Steganopoda (=Totipalmatæ, Dysporomorphæ): all the toes united by a membrane. Longipennes: hind toe free; wings long [from these Procellariidæ were separated as Tubinares; Aptenodytidæ are the Spheniscomorphæ of Huxley]. Pygopoda (=Brevipennes, Urinatores): hind toe sometimes absent; wings very short. *Phœnicopterus* is placed by Huxley between the anserine birds and the storks and herons, under the name of Amphimorphæ. Garrod makes it a subfamily of bustards. Its webbed feet is said by Owen to be an artificial character.

LAMELLIROSTRES.

Phœnicopteridæ.

Phœnicopterus (Flamingo).

Anatidæ.

Cygnus (Swan).
Anser (Goose).
Bernicla.
Cereopsis.
Plectropterus.
Tadorna.
Cairina.
Rhynchaspis (Sheldrake).
Anas (Duck, &c.).
Querquedula (Teal, &c.).
Aix=Dendronessa.
Harelda.
Fuligula (Pochard).
Clangula.

Somateria (Eider).
Œdemia (Scoter).
Erismatura.
Biziura.

Mergidæ.

Mergus (Goosander, Smew, &c.).

STEGANOPODA.

Pelecanidæ.

Sula=Dysporus (Gannet).
Phalacrocorax=Carbo=Haliæus (Shag, &c.).
Plotus (Darter).
Pelecanus (Pelican).

Tachypetidæ.

Tachypetes (Frigatebird).

Phaëthontidæ.

Phaëthon (Tropicbird).

LONGIPENNES.

Procellariidæ.

Thalassidroma (Petrel).
Æstrelata.
Prion.
Procellaria (Fulmar).
Puffinus (Shearwater, Mollymock).
Halodroma.
Diomedea (Albatross).

Laridæ.

Lestris (Skua).
Larus (Gull).

Sternidæ.	*Alcidæ.*	Chimerina.
Sterna (Tern).	Alca (Razor-bill).	*Aptenodytidæ*
Anous (Noddy).	Uria (Guillemot or	(Penguins).
Rhynchops.	Murre).	Spheniscus.
	Grylle.	Eudyptes = Catar-
PYGOPODA.	Mergulus.	rhactes.
	Mormon = Frater-	Aptenodytes.
Colymbidæ.	cula (Puffin).	
Colymbus (Diver).	Phaleris.	*Cnemiornis.
Podiceps (Grebe).		

Order X. ACCIPITRES.

RAPTORES. RAPTATORES. ÆTOMORPHÆ.

Bill arched, strongly hooked at the tip; a cere at the base in which the nostrils are placed. Legs stout, the inner toe only turned backwards; claws hooked, sharp, and partially retractile.

The strongly hooked bill is armed with a lateral tooth in the carnivorous species, but in the carrion-feeders it is obsolete or wanting. In some of the eagles (*Aquila, Pandion, Haliaetus*) the sternum is entire. In the Cathartidæ only are the claws blunt and comparatively straight. There are two carotids. The lower larynx, worked by only one pair of intrinsic muscles, is present in all except the Cathartidæ. *Gypogeranus* is remarkable for the length of the tarso-metatarsus. The tarsi and toes in this order are generally reticulated, and sometimes clothed with feathers. Excepting the owls, the females are always larger than the males.

In the owls the eyes are frontal, and there is a large circle of feathers around each; the ear is well developed, and often provided with an operculum. The iris is enlarged to allow a greater dilatation of the pupils that more light may enter the eye. The outer or fourth toe is reversible. Their plumage is peculiarly soft and downy.

It is only among the vultures that social species are found. Hawks and owls are solitary birds, building rude nests in almost inaccessible situations, the latter in holes of trees, laying few eggs.

In this order we have the condor (*Sarcoramphus gryphus*), king-vulture (*Sarcoramphus papa*), turkey-buzzard (*Cathartes aura*), lammergeyer (*Gypaëtus barbatus*), goshawk (*Astur palumbarius*), sparrow-hawk (*Accipiter nisus*), merlin (*Falco æsalon*), kite (*Milvus regalis*), buzzard (*Buteo vulgaris*), osprey (*Pandion haliaetus*), eagle (*Aquila chrysaetos*), falcon (*Falco peregrinus*), kestrel (*Cerchneis tinnunculus*), barn-owl (*Strix flammea*), great eagle-owl (*Bubo maximus*), screech-owl, tawny owl, or ivy-owl (*Syrnium aluco*), snowy owl (*Nyctea nivea*), and burrowing-owl (*Athene cunicularia*).

The Accipitres are sometimes divided into diurnal and nocturnal. Huxley's Gypaëtidæ comprise Vulturidæ and Falconidæ. *Dicholophus* is a very aberrant form, frequently classed with the Grallæ; by Sundevall it is placed with *Polyborus* to form his Polyborinæ, which, with "Cartharteæ," constitute his fourth cohort of Accipitres (Necroharpages). *Harpagornis* is a large extinct New-Zealand form.

Cathartidæ (Vultures of the New World).

Sarcoramphus.
Cathartes.

Vulturidæ (Vultures of the Old World).

Gypaëtus.
Neophron.
Gyps.
Vultur.

Falconidæ.

Polyborus.
Ibycter.
Gymnogenys.
Astur (Goshawk).

Accipiter=Nisus (Sparrow-hawk).
Nauclerus.
Elanus.
Milvus (Kite).
Pernis (Honey-buzzard).
Buteo (Buzzard).
Circus (Hen-harrier).
Pandion (Osprey).
Circaëtus.
Harpyia.
Aquila (Eagle).
Haliaetus.
Falco (Falcon).
Cerchneis (Kestrel).

Strigidæ (Owls).

Athene.
Surnia.
Nyctea.
Asio=Otus.
Scops.
Bubo.
Smilonyx=Ketupa.
Nyctea.
Syrnium.
Strix.

Gypogeranidæ.

Gypogeranus=Serpentarius.

Dicholophus=Cariama.

*Harpagornis.

Order XI. PSITTACI.

PREHENSORES. PSITTACOMORPHÆ.

Bill short, stout, arched, hooked at the tip; a cere at the base. Tongue fleshy. Tarsi short, reticulate. Outer and inner toes turned backwards.

The upper mandible is articulated with the frontal bones by a complete hinge-joint; the nostrils are pierced in the cere. The tongue is unusually soft and fleshy, and has a brush at the end in Trichoglossidæ. The inferior larynx has three pairs of muscles. The clavicles are comparatively weak, and are sometimes absent.

The sternum is either perforated or entire. There are two carotids. The gall-bladder is generally absent.

In the fœtal parrakeet the margins of the bill are beset with
tubercles, under each of which is a gelatinous pulp, like that of
a tooth.

The bill assists in climbing, and the feet are used as hands, a
peculiarity possessed only by these birds and the goatsuckers.

These birds are monogamous, living in society, and forming
their nests in holes of trees. The females have the same rich
colours as the males. They feed mostly on seeds and fruits, and
scream horribly. They are the only real climbers (hand-over-
hand) among birds. Some, as the grey parrot, are known to live
to a great age.

The grey parrot (*Psittacus erithacus*), ground-parrakeet (*Pezo-
porus formosus*), canary parrakeet (*Melopsittacus undulatus*), the
love-bird (*Agapornis Swinderiana*), white cockatoo (*Plyctolophus
sulphureus*), the macaws (*Macrocercus*, sps.), and the green parrots of Brazil (*Chrysotis*, sps.) belong to this order. *Strigops* is
an owl-like, nocturnal bird of New Zealand, feeding on roots, &c.
It has a keelless breast-bone.

There are about 430 known species.

Strigopidæ.	Pezoporus.	Microglossus.
Strigops.	Platycercus.	Plyctolophus = Ca-
	Palæornis.	catua (Cockatoo).
Trichoglossidæ.	Conurus.	
	Henicognathus.	*Psittacidæ* (Parrots).
Nestor.	Macrocercus = Ara =	
Eos.	Sittace (Macaw).	Coryllis.
Lorius = Domicellus		Agapornis = Psitta-
(Lory).		cula.
Coriphilus.	*Plyctolophidæ.*	Brotogerys.
Trichoglossus (Par-	Nasiterna.	Chrysotis.
rakeet).	Nymphicus = Calop-	Pionus = Deroptyus.
	sitta.	Eclectus.
Macrocercidæ.	Licmetis.	Dasyptilus.
Melopsittacus	Calyptorhynchus.	Psittacus.
= Nanodes.		

Order XII. ODONTORMÆ.

Jaws with teeth in separate sockets. "Vertebræ biconcave.
Sternum with a keel. Wings well developed."

Extinct "Reptilian birds" found in the Cretaceous shales of
Kansas. The species are supposed to have been carnivorous and

aquatic. *Ichthyornis* was about the size of a pigeon. *Odonto-pteryx toliapicus*, from the Isle of Sheppey, regarded as*most nearly related to Natatores, has bony processes of the jaws equivalent to teeth.

*Ichthyornis.   *Apatornis.

*Odontopteryx.

Order XIII. ODONTOLCÆ.

Jaws with the teeth in continuous grooves. "Vertebræ as in birds. Sternum without a keel. Wings rudimentary."

Extinct birds of large size, contemporaneous with the last. They are also supposed by Marsh, to whom we owe all we know of this and the preceding order, to have been carnivorous and aquatic. *Hesperornis* he compares with the cassowaries and the penguins.

*Hesperornis.   *Lestornis.

Order XIV. SAURURÆ.

URoIoNI.

Tail composed of numerous free vertebræ, each carrying a single pair of feathers. Sternum rudimentary. Metacarpal bones not anchylosed.

The first known specimen of the remarkable form (*Archæopteryx lithographica*) which alone constitutes this order was found in 1861, in the lithographic limestone of Solenhofen, near Munich. It was about the size of a pigeon, and was supposed to have been web-footed and a fish-eater. A more perfect specimen has recently been discovered. Two small conical teeth remained in the upper jaw; there was no appearance of a furcula, and the sternum was "reduced to zero." The manus resembles that of a tridactyle lizard. According to Vogt, its only bird-characters were its feathers and feet. The tail, always a variable organ, shows in this animal the persistency of what is now an embryonal character.

The *Archæopteryx* is what Huxley has called an "intercalary type," that is, not representing the *direct* passage from one group to another. It would seem to lie somewhere between the gulls (or perhaps falcons) and the extinct reptilian form *Compsognathus*.

*Archæopteryx.

Class V. MAMMALIA.

Vertebrate, warm-blooded, viviparous animals, more or less covered with hair, breathing by lungs, the females having mammary glands (rudimentary in the males). A corpus callosum.

Mammals are the only animals that suckle their young, and which, before birth, are nourished by a placenta. There are two types of placentas: in the one the uterus produces a "decidua," a modification of its mucous membrane, between which and the chorion a placenta is gradually formed; in the other there is no decidua, the uterus simply becoming more vascular and forming a union with the chorion, which is dissolved at parturition.

It is only in the brain of mammals that we find a greater or anterior commissure [corpus callosum] uniting the two hemispheres of the cerebrum, and a lesser commissure [pons Varolii] those of the cerebellum; but they are small or wanting in the Monotremata and Marsupialia.

The skull is articulated to the atlas by double condyles, and the lower jaw to the temporal bone without the intermediation of an os quadratum, as in birds and reptiles.

The teeth are fixed in distinct sockets, and "usually consist of hard unvascular dentine, defended at the crown by an investment of enamel, and everywhere surrounded by a coat of cement." The typical number is forty-four. In some mammals the teeth are permanent [monophyodont], in others the first-developed teeth are succeeded by another set [diphyodont].

The heart has two auricles and two ventricles. The abdominal are separated from the thoracic viscera by the midrib, or diaphragm. Unlike birds and reptiles, the kidneys are composed of two parts—cortical and medullary; one kidney is always placed higher than the other. There are two ovaries, but in the Monotremata the right one is rudimentary. A membrana nictitans is present in nearly all mammals except man and monkeys, but in whom it is represented by the plica semilunaris. In the mole and *Spalax typhlus* the eyes are obsolete, but in the former they are well developed in the embryo. Most mammals have an external ear (concha). The young are often born blind; they acquire their sight in from eight to fifty days; others see at once, and are able to accompany their mother a few hours after birth.

Among the many systems of classification of the Mammalia,

the most remarkable is that which Professor Owen laid before
the Linnean Society in 1857, based on four leading modifications
of the brain. In the "first and lowest primary group or subclass"—Lyencephala—the cerebral hemispheres are smooth and
without folds, leaving the olfactory ganglions, cerebellum, and
optic lobes more or less exposed [Monotremata, Marsupialia].
(2) Lissencephala : cerebral hemispheres with few folds ; olfactory lobes and cerebellum exposed, but a corpus callosum present
[Rodentia, Chiroptera, Insectivora, Bruta].   (3) Gyrencephala :
hemispheres folded into more or less numerous "gyri," and
extending more or less over the cerebellum and olfactory lobes
[Cetacea, Ungulata, Quadrumana].   (4) Archencephala : hemispheres more folded, overlapping the olfactory lobes and cerebellum [Man only].   The last character was said to be peculiar
to the genus *Homo*, and also "equally peculiar." were the ' posterior horn of the lateral ventricle' and the 'hippocampus minor,'
"which characterize the hind lobe of each hemisphere." But
these characters are now known to exist in the ourang-outang
and all the higher Quadrumana.

Milne Edwards and Alphonse Milne-Edwards have given a
classification of the Mammalia as below :—

Première sous-classe Mammifères normaux.
    Phalange des Hématogénètes.
        Légion des Micrallantoïdés.
            Cohorte des Primates (Bimanes, Quadrumanes).
            Cohorte des Plébéiates (Chiroptères, Insectivores, Ron-
                geurs).
        Légion des Mésallantoïdés (Carnivores, Pinnés ou Am-
            phibies).
    Phalange des Hyraciens.
    Phalange des Proboscidiens.
    Phalange des Mégallantoïdiens (Pachydermes solidongulés,
        Pachydermes bisulques, Camélides, Tragulides, Pécorides).
    Phalange des Édentés.
Deuxième sous-classe Pinnifériens ou Mammifères Piscif rmes
    (Sirénides, Cétacés).
Troisième sous-classe (Marsipiaux, Monotrèmes).

The Mammalia are divisible into Non-placentals and Placentals.
In the former De Blainville includes two subclasses—Ornitho-
delphia (=Monotremata) and Didelphia (=Marsupialia), while
the latter corresponds to his third subclass—Monodelphia.

Huxley, also considering that the placenta affords the "best characters" for classification, divides them as follows :—*A discoidal deciduate placenta* : Primates, Insectivora, Chiroptera, Rodentia. *A zonary deciduate placenta* : Carnivora, Proboscidea, Hyracoidea. *A non-deciduate placenta* : Ungulata, Cetacea. [Placenta of Sirenia unknown.] *A variable placenta* : Edentata.

The classification adopted here is substantially the one now in use, only that the Pinnipedia are sometimes combined with the Carnivora, and the Hyracoidea are not always disunited from the Proboscidea, or Sirenia from Cetacea. The Lemurs also, under the name of "Prosimii," or "Prosimiæ," are by some separated from the Quadrumana.

Without a placenta [Implacentalia].
  No abdominal pouch .... MONOTREMATA.
  An abdominal pouch in the female............... MARSUPIALIA.
With a placenta [Placentalia].
  With hind legs.
    Hind legs free.
      No thumb opposable to the fingers.
        Unguiculate [claws on the upper part
        of the toes only].
          Claws small or moderate.
            No canines............................ RODENTIA.
            With canines.
              Fore limbs formed for flight ... CHIROPTERA.
              Fore limbs normal.
                Canines small .... INSECTIVORA.
                Canines large .... CARNIVORA.
          Claws very large .... BRUTA.
        Ungulate [claws enclosing the toes, *i.e.*
        hoofs].
          Placenta deciduate.
            Incisors tusk-formed .... PROBOSCIDEA.
            Incisors normal...................... HYRACOIDEA.
          Placenta non-deciduate .... UNGULATA.
      A thumb opposable to the fingers.
        Teeth uneven, interrupted .... QUADRUMANA.
        Teeth even, uninterrupted .... BIMANA.
    Hind legs fin-like .... PINNIPEDIA.
  No hind legs, a horizontal tail-fin.
    Nostrils on the muzzle........................ SIRENIA.
    Nostrils on top of the head .... CETACEA.

As no linear arrangement of the Mammalia can be quite satisfactory, an attempt is here made to show the affinities of the orders in a diagrammatic form. No species of Bruta, however, —the lowest of the placental Mammalia, and a fragmentary order—appears to approximate to any known Marsupial. The passages from the latter to the Rodentia is made by the wombats, and to the Insectivora by the smaller opossums. The affinities of the remainder are not doubtful.

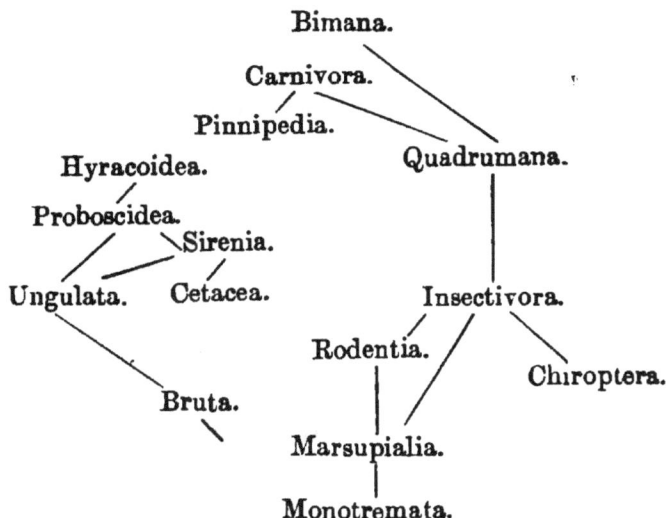

Bimana.

Carnivora.

Pinnipedia.

Quadrumana.

Hyracoidea.

Proboscidea.

Sirenia.

Ungulata.   Cetacea.

Insectivora.

Rodentia.

Chiroptera.

Bruta.

Marsupialia.

Monotremata.

According to Schmarda there are above 2300 recent and 300 extinct species of mammals.

Order I. MONOTREMATA.

Ornithodelphia.

A common cloaca. Jaws without true teeth. No marsupial pouch. Mammary gland without a nipple. No corpus callosum.

The face is prolonged into the form of a bird-like beak, covered by a smooth skin, the mouth at the extremity and without fleshy lips, although they are for a time manifest in the young.

Another inferior character, like the common cloaca, is that the coracoid bone extends from the scapula to the sternum; they have also an epicoracoid or interclavicle, and an episternum, as in the lizards. Marsupial bones [wrongly so called, *Huxley*] are present, but there is no marsupial pouch. There are no external ears. The corpus callosum is wanting, and the optic lobes are simple or undivided.

Of the two genera, *Echidna* is terrestrial, with beak-like jaws, and a small mouth at the end, a long slender tongue, and a body covered with spines, and with an exceedingly short or almost obsolete tail; the male has a perforated spur. There are two Australian species, and one or two from New Guinea. It is doubtful if the species are oviparous or ovoviviparous. *Ornithorhynchus* is aquatic, with a flat duck-like beak, short tongue, mole-like fur, and a broad flat tail of moderate length. A single Australian species is known—the duck-mole or water-mole (*Ornithorhynchus paradoxus*). It is a good swimmer and constructs long burrows in the banks of rivers; it feeds like a duck, sucking up its prey from the mud.

Echidnidæ.	*Ornithorhynchidæ.*
Echidna = Tachyglossus.	Ornithorhynchus = Platypus.

Order II. MARSUPIALIA.

DIDELPHIA.

An abdominal pouch in the female. True teeth of two or three kinds. No common cloaca. No placenta.

The abdominal pouch [marsupium] is supported by the marsupial bones, which are ossifications of the inner tendon of the external oblique muscle. It is into this pouch that the prematurely born offspring is transferred, the young animal remaining suspended from the nipple, and so helpless as to be unequal to the muscular effort of sucking. The mother, however, has the mammary gland provided with a cremaster muscle, by which she is able to force her milk into the mouth of her pendent young. The marsupial bones occur also in the males, but without the pouch. The coracoid, as in the higher Mammalia, forms part of the scapula, and is not attached to the sternum. The corpus callosum is very small or wanting. The size of the brain is $\frac{1}{25}$ in *Petaurista pygmæa*, and $\frac{1}{100}$ in *Macropus major*.

The fore and hind limbs are somtimes connected by an extension of the integument from the side, as in *Petaurus* and *Acrobata*. In *Chironectes*, the only aquatic form, the feet are webbed. The opossums have on their hind feet a thumb opposable to the digits, as in the Quadrumana.

The affinity of the Implacentalia to the Sauria is shown in parts of the skeleton. One reptilian character is that the bones of the skull remain in a state of permanent separation.

The Marsupials are amongst the oldest known mammals, and some were of very large size. *Diprotodon australis* had a skull three feet in length. Several species are found in the British Mesozoic formations.

The recent species are mostly Australian, but they were once common in Europe. Didelphidæ are American. Owen has divided them into five "tribes," all of which are represented among the monodelphous Mammalia. Reversing the descending order, these tribes are:—

RHIZOPHAGA.—Two scalpriform incisors in both jaws; no canines; short cæcum.

POËPHAGA.—Anterior incisors large and long in both jaws; canines in the upper jaw only, or wanting; a long cæcum.

CARPOPHAGA.—Anterior incisors large and long in both jaws canines inconstant; a long cæcum.

ENTOMOPHAGA.—Three kinds of teeth in both jaws; a cæcum.

SARCOPHAGA.—Three kinds of teeth in both jaws; no cæcum.

RHIZOPHAGA.

Phascolomyidæ.
Phascolomys (Wombat).

———

*Diprotodon.

POËPHAGA.

Macropodidæ.
Dendrolagus.
Hypsiprymnus (Kangaroo-rat).
Macropus = Halmaturus (Kangaroo).

———

Hypsiprymnodon.

CARPOPHAGA.

Phascolarctidæ.
Phascolarctos (Koala).

Phalangistidæ.
Cuscus.
Phalangista.
Acrobata.
Petaurus.

ENTOMOPHAGA.

Peramelidæ.
Chœropus.
Perameles (Bandicoot).

———

*Thylacoleo.

Tarsipes.

Didelphyidæ.
Didelphys (Opossum).
Chironectes.

SARCOPHAGA.

Dasyuridæ.
Thylacinus.
Dasyurus.
Myrmecobius.
Phascogale.

———

*Phascolotherium.

———

*Microlestes.

Order III. BRUTA.

EDENTATA.

Teeth small, or sometimes wanting; no median incisors; no canines. Feet with long and strong claws.

There are no teeth in *Manis* and *Myrmecophaga*; but in *Dasypus gigas* there are about 100, all molars. There is no second set, and they are without enamel and destitute of roots. The tongue in the toothless species is long and extensible. In Dasypodidæ the skin of the upper parts is covered with bony plates [scutes], in *Manis* with scales; hair is also present. There are two pectoral mammæ, to which sometimes a pair of inguinal or abdominal is added. The stomach is musculo-tendinous, a set-off for the low power of mastication. There is no cæcum in *Manis*, and there are two cæca in *Dasypus*. *Cholœpus* has twenty-three pairs of ribs, the largest number known among the Mammalia.

The placenta is variable but mostly non-deciduate. There is only one young at a birth, except in the armadillos. In the great anteater the young remains with the mother for a year, and is carried on her back.

The Bruta are very slow in their movements, have a very feeble cry, or are mute. The sloths are arboreal; the remainder are mostly burrowing; therefore they have unusually powerful fore limbs and stout clavicles. The extinct species were of large size, and appear to have been numerous; South America then, as now, was the region most affected by them.

To this order belong the great anteater (*Myrmecophaga jubata*), armadillo (*Dasypus sexcinctus* and *D. peba*), and sloth (*Bradypus torquatus*). Two species of *Orycteropus* are the only African representatives of the family; while *Manis* represents it in Asia. *Manis*, *Myrmecophaga*, and *Orycteropus* form the family Vermilinguia of some writers. Dasypodidæ are synonymous with Cingulata or Loricata. Tardigrada are the Bradypodidæ, and Gravigrada the Megatheriidæ.

Manidæ.	*Myrmecophagidæ.*	*Orycteropodidæ.*
Manis (Pangolin).	Myrmecophaga (Anteater).	Orycteropus (Ground-hog).
Macrotheriidæ.	Cyclothurus.	
*Macrotherium.	*Glossotherium.	*Dasypodidæ.*
*Ancylotherium.		Chlamyphorus.

Tolypeutes.	*Megatheriidæ.*	*Bradypodidæ*(Sloths).
Dasypus(Armadillo).	*Megalonyx.	Bradypus=Acheus.
Glyptodontidæ.	*Megatherium.	Cholopus.
*Glyptodon.	*Mylodon	
	*Sphenodon.	

Order IV. RODENTIA.

GLIRES. ROSORES.

Two long, incurved incisors in each jaw, remote from the molars; no canines. Hallux not opposable to the digits. Clavicles generally present.

The molar teeth are few in number, and are transversely penetrated by the enamel; they have often no roots, or the roots are tardily produced. The incisors, separated by a wide interval from the molars, are bevelled off on the inner surface from attrition; but they continue to grow from the base through life, and thus they preserve a uniform length. In early life in hares and rabbits there are six upper incisors, but four of them are deciduous. There are, however, in the adult two supplementary teeth behind the upper ones. The upper lip is sometimes divided. The fore feet are frequently used as hands; the hind feet in a few species, like the jerboa, are of great length, chiefly owing to the extreme development of the metatarsal bone. The mammæ vary from two to ten, and are pectoro-abdominal or entirely abdominal. The stomach is simple; the cæcum is of large size, but is absent in the dormice, and the intestines are very long.

Some of the Rodentia build nests; they are very prolific; many, especially in northern climates, hibernate. In a few of those that lead a subterranean life the eyes are reduced to a small bulb. The brain is smooth and without convolutions. The orbits are incomplete, not being separated from the temporal fossæ. Clavicles may or may not be present.

The greater part of the Rodentia feed on vegetable matter of one kind or another, but some are omnivorous. About 700 species are known, 100 of these are found in Europe.

Among the members of this order are, the hare (*Lepus timidus*), rabbit (*Lepus cuniculus*), chinchilla (*Chinchilla lanigera*), guinea-pig(*Cavia aperea*), porcupine (*Hystrix cristata*), lemming (*Myodes lemmus*), black rat (*Mus rattus*), the brown rat, introduced, and now the commonest of the two (*Mus decumanus*), mouse (*Mus musculus*), harvest-mouse (*Mus messorius*), wood-mouse (*Mus syl-*

vaticus), hamster (*Cricetus frumentarius*), jerboa (*Dipus sagitta*), dormouse (*Myoxus avellanarius*), water-rat or vole (*Arvicola amphibius*), field-mouse (*Arvicola agrestis*), beaver (*Castor fiber*), prairie-dog (*Cynomys ludovicianus*), marmot (*Arctomys marmota*), and squirrel (*Sciurus vulgaris*).

Alston divides the Rodentia into three suborders, according to the characters of the teeth; one of these is subdivided into three sections: these are indicated in the following list:—

HEBEDIDENTATA.

Mesotheriidæ.

*Mesotherium.

DUPLICIDENTATI.

Leporidæ.

Lagomys.
Lepus (Hare, Rabbit).

SIMPLICIDENTATI.

Hystricomorpha.

Caviidæ.

Dasyprocta (Agouti).
Cœlogenys.
Dolichotis.
Cavia (Cavy).
Hydrochœrus.

Dinomyidæ.

Dinomys.

Chinchillidæ.

Lagostomus.
Lagotis = Lagidium.
Chinchilla = Eriomys.

Octodontidæ.

Canromys.

Echimys.
Loncheres.
Myopotamus.
Habrocoma.
Octodon.
Ctenomys.
Petromys.
Pectinator.
Ctenodactylus.

Hystricidæ.

Synetheres = Cercolabes.
Erethizon.
Atherura.
Hystrix (Porcupine).

Myomorpha.

Spalacidæ.

Myospalax.
Bathyergus.
Georhychus.
Spalax.

Dipodidæ.

Pedetes.
Scirtetes = Alactaga.
Jaculus.
Dipus (Jerboa).

Muridæ.

Siphneus.

Myodes = Lemmus (Lemming).
Arvicola (Vole).
Fiber.
Hypudæus.
Hydromys.
Psammomys.
Gerbillus.
Meriones.
Hesperomys.
Hapalotis.
Dendromys.
Mus (Mouse, Rat).
Sminthus.

Lophiomys.

Cricetidæ.

Saccostomus.
Cricetus (Hamster).

Saccomyidæ.

Geomys.
Saccomys.

Myoxidæ.

Myoxus (Dormouse).
Graphiurus.

Sciuromorpha.

Anomaluridæ.

Anomalurus.

Castoridæ.	Spermophilus.	*Haplodontiidæ.*
Castor (Beaver).	Tamias.	Haplodontia.
	Pteromys (Flying-	
Sciuridæ.	Squirrel).	*Ischyromyidæ.*
	Xeros.	
Arctomys (Marmot).	Sciurus (Squirrel).	*Ischyromys.
Cynomys.		

Order V. CHIROPTERA. (Bats.)

VOLITANTIA.

Fore limbs with four prolonged ulnar digits, united by an extension of the integument. Three kinds of teeth. One or two pairs of pectoral mammæ.

The pollex (or thumb) is free, and not prolonged as are the digits, and is alone furnished with a claw or nail. A membranous skin extending from the lower part of the neck to the extremities of the digits, and usually continued to the tail, forms a powerful organ of flight [patagium]. The bones contain no air-cells as in birds, nor is there any mesial crest to the sternum. The eyes are small; but the ears are often very large and, as well as the membrane of the wings, are probably subservient to the sense of touch. The nose, also, is sometimes furnished with peculiar membranous expansions. In the tongue of the Phyllostomidæ there is a peculiar disposition of the terminal papillæ, enabling it to act as an organ of suction. These bats, it is well known, suck the blood of other animals, man included. The vampyre (*Vampyrus spectrum*) is one of the worst.

Bats are crepuscular or nocturnal and hibernate in cold climates. Except Pteropodidæ, which are exclusively frugivorous, they are nearly all insect-feeders. When at rest, they suspend themselves by one of their thumbs, or hang head downwards, holding on by their hind feet.

The female bat brings forth one or two at a birth, which she carries about with her. The young, in some species at least, are born blind and destitute of hair.

Six families are recognized; but the genera and "subgenera" are very numerous. Dobson divides them into two suborders—Megachiroptera and Microchiroptera—the large and the small bats respectively. There are some 500 species, of which about 17 are British. Our common bat is *Vespertilio pipistrellus*; the long-eared bat (*Plecotus auritus*) is also common; *V. murinus* is only a very rare straggler.

Phyllostomidæ.
Glossophaga.
Anura.
Vampyrus.
Phyllostoma.
Stenoderma.
Carollia.
Artibeus.
Mormops.

———

Desmodus.

Noctilionidæ.
Emballonura.

Taphozous.
Rhinopoma.
Noctilio.
Dysopes = Molossus.
Mystacina.

Vespertilionidæ.

Atalapha.
Plecotus.
Synotus.
Vesperugo = Scoto-
  philus.
Vespertilio.

Megadermatidæ.

Nycteris.
Megaderma.

Rhinolophidæ.

Rhinolophus.
Phyllorhina.
Harpyia.
Macroglossus.
Cynopterus.
Pteropus.
Cœlops.

Order VI. INSECTIVORA.

Plantigrade. Three kinds of teeth; incisors short, simple. Five toes, furnished with claws. Clavicles present.

The canine teeth are small and the molars are studded with small tubercles, fitting them more for grinding than for tearing their food. The limbs are short and feeble, except in the Macroscelidæ &c. The mammæ are abdominal. The placenta is deciduate and discoidal. The brain is without convolutions. A gall-bladder is always present.

In the mole the eye is very small and nearly covered by the skin; but in the embryo it is as well developed as usual. The hedgehog has a powerful cutaneous muscle attached to the integument, by which it is enabled to roll itself into a ball or to erect its spines. This power is not possessed by its allies. In the Soricidæ the lower incisors become anchylosed to the jaw-bone, a reptilian character not occurring in any other mammal.

· The Insectivora are mostly nocturnal and subterranean, and many hibernate. The Tupaiidæ are arboreal, and live on fruit as well as on insects. A few are aquatic or semiaquatic.

In this order we have the hedgehog (*Erinaceus europæus*), two species of shrews generally confounded together, both common (*Sorex tetragonurus* and *Sorex rusticus*), [*Crocidura aranea* has not been found in England], water-shrew (*Crossopus fodiens*), and mole (*Talpa europæa*). *Galeopithecus* is placed here by Huxley, but by Claus and Schmarda it is retained among the lemurs. With some of the characters of the latter, it has also a tegumentary membrane extending from the neck to the tail, and embracing the legs.

Talpidæ.

Urotrichus.
Scalops.
Condylura.
Chrysochloris.
Talpa (Mole).

Soricidæ.

Myogale=Mygale
 (Desman).
Pachyura.

Crocidura.
Crossopus.
Sorex (Shrew).

Macroscelidæ.

Rhynchocyon.
Macroscelides.

Tupaiidæ.

Tupaia=Cladobates.
Hylomys.

Erinaceidæ.

Centetes.
Echinogale.
Erinaceus (Hedge-
 hog).
Gymnura.
————
Solenodon.

Potamogalidæ.

Potamogale.

Galeopithecidæ.
Galeopithecus (Flying lemur).

Order VII. HYRACOIDEA.

LAMNUNGUIA. GLIRIFORMIA.

Two long curved incisors in the upper jaw; no canines; molars 12-14 in each jaw. Toes with flattened nails, the hind toe with a claw. Mammæ four inguinal, two pectoral.

The lower incisors are straight; they are four in number, the upper are only two. There are no clavicles. The stomach is complex [simple, *Huxley*], the cæcum "very large," and there is no gall-bladder. As in the preceding order and in the Carnivora, there is a zonary deciduate placenta.

This is a very small group, with four or perhaps five species. One of them, *Hyrax syriacus*, is the coney of Scripture; another species (*Hyrax capensis*) is the "badger" of the Cape colonists. It was at one time considered to be a Rodent. Very recently it has been combined with the elephant to form the order "Chelophora" [*v. Koch*].

Hyracidæ.
Hyrax.

Order VIII. PROBOSCIDEA.

Two tusk-like incisors in the upper jaw; no canines; molars few. Nose prolonged into a proboscis, with the nostrils at the end. Mammæ two, pectoral.

In some of the extinct species there are incisors in the lower jaw. In the elephant, the only recent genus, the two large per-

manent tusks, composed entirely of dentine, are preceded by two
deciduous smaller ones. The structure of the molars is exceedingly complex, but there is only one molar at a time in
each jaw. The proboscis is a long flexible organ, known as
the trunk; it is terminated by a thumb-like appendage, and
encloses a double tube, between which and the integument is
a thick layer of muscular substance. The cranial cavity is very
small, the immense size of the head being due to the enormous
development of the frontal sinuses. This peculiarity depends on
the necessity of supporting the tusks, and of affording sufficient
surface for the attachment of muscles to work the trunk. There
are no clavicles. The stomach is simple, the intestines very long
and voluminous, and the cæcum of large size. The placenta is
zonary and deciduate.

The elephant is herbivorous; every thing it eats is put into its
mouth by its trunk. There are two species—*Elephas indicus*,
with smaller ears and the males alone with well-developed tusks;
and *E. africanus* (separated as a genus—*Loxodon*—by Gray), with
large ears and well-developed tusks in both sexes; the enamel is
also markedly different. A third species, the mammoth (*E. primi-
genius*), is now extinct, but it was once common in England.
Fossil species are somewhat numerous; their remains have been
found in all parts of the world; one of them, *Dinotherium*, has
been regarded as a dugong and as a Marsupial. Like *Mastodon*
and the extinct elephants, it belonged to the Miocene period.

Elephantidæ. *Dinotheriidæ.*
*Mastodon. *Dinotherium.
Elephas (Elephant).

Order IX. UNGULATA.

Toes never more than four, protected by hoof-like nails, and
incapable of grasping. No clavicles. Placenta non-deciduate.

There are two sets of enamelled teeth; canines are not often
present; the lower incisors are sometimes wanting, and the upper
incisors in most of the ruminants are replaced by a callosity of
the gum. There are generally six molars on each side of either
jaw; these teeth are deeply penetrated by the enamel, and are of a
massive character. In the pig family and musk-deer the canines
are in the form of tusks.

Many of the Ungulata are furnished with horns. These are of
various kinds:—a solid prolongation of bone, covered by the skin
and persistent, as in the giraffe; a solid outgrowth of the frontal

bones, confined (except in the reindeer) to the males, frequently branched and always deciduous, as in the deer family; neither of these are true horns; the latter are, indeed, more generally called *antlers*. The horns in the ox and antelope families are hollow and persistent, consisting of a bony core covered by a corneous case. These are known as the Cavicornia. The nasal horns of the Rhinoceroses are composed of concrete hairy fibres, closely compacted and fixed on a bony protuberance.

In the ruminant families the feet are bisulcate, with two supplementary hoofs in some, placed above and at the back of the foot. Camelidæ (=Tylopoda, Phalangigrada) have two toes, callous beneath, the hoofs scarcely more than rudimentary. None of the Ungulata have clavicles.

The stomach is divided into four compartments—the two upper essentially dilatations of the œsophagus, the third or, rather, perhaps the fourth, being the true stomach. In the stag family there is a sebaceous gland [lachrymal sinus] in front of the eye, secreting a disagreeable waxy substance.

The placenta is diffused in the camels and in non-ruminants; and is cotyledonary in the true ruminants.

The Ungulata pretty nearly include the old orders Belluæ, Pecora, Ruminantia, Pachydermata, and Solipeda or Solidungula. They were divided by Owen into two orders—Artiodactyla and the Perissodactyla (the former with paired, the latter with unpaired toes); these are now reduced to suborders.

The animals most important to man are included in the Ungulata. Among them are the camel or dromedary (*Camelus dromedarius*), the wild camel. now domesticated (*Camelus bactrianus*), alpaca (*Auchenia pacos*), lama (*Auchenia lama*), giraffe (*Camelopardalis giraffa*), stag (*Cervus elaphus*), wapiti (*Cervus canadensis*),roebuck (*Cervus capreolus*), reindeer (*Cervus tarandus*), elk (*Alces malchis*), fallow-deer (*Dama platyceros*), musk-deer (*Moschus moschiferus*), saiga (*Colus tataricus*), gazelle (*Antilope dorcus*), water-buck (*Cervicapra ellipsiprymnus*), chamois (*Rupicapra tragus*), ibex (*Capra ibex*), goat (*Capra hircus*), sheep (*Ovis aries*), wild sheep or moufflon (*Ovis musimon*), bighorn (*Ovis montana*), musk-ox (*Ovibos moschatus*), ox (*Bos taurus*), buffalo (*Bos bubalis*), bison or auroch (*Bos bison*) [the buffalo of North America is scarcely distinct], Cape buffalo (*Bos caffer*), zebu (*Bos indicus*), and yak (*Poëphagus grunniens*). Among the non-ruminants are the hippopotamus (*Hippopotamus amphibius*), pig (*Sus scrofa*), wild boar (*Sus aper*), rhinoceros [about five species; the best known perhaps is *Rhinoceros bicornis*, but an extinct species (*R. tichorhinus*) was once very common in England], tapir (*Tapirus americanus*), horse (*Equus caballus*), ass (*Equus asinus*), and zebra (*Equus zebra*).

ÁRTIODACTYLA.—Two or four toes. Stomach complex; cæcum small, wanting in *Hippopotamus*.

Ruminants subject their food to a second mastication, bringing it up from the stomach into which it had passed. They have the "cloven foot"—a median pair of hoofed toes, and generally two behind.

The musk-deer, according to Flower, is a low and doubtful type of Cervidæ. It has until recently been placed with *Tragulus*, which is now considered to constitute a distinct family. The musk of commerce is secreted in an abdominal pouch found only in the males.

Many extinct genera "once linked together the now broken series "of this suborder.

"NON-RUMINANT"

(=Bunadontia).

Hippopotamidæ.

Hippopotamus.

Suidæ.

Phacochœrus.
Dicotyles (Peccary).
Sus (Pig).
*Chœropotamus.
*Anthracotherium.
*Hyopotamus.

Anoplotheriidæ.

*Anoplotherium.
*Dichobune.
*Dichodon.

"RUMINANT"

(=Selenodontia).

Camelopardalidæ.

Camelopardalis (Giraffe).
*Sivatherium.

Camelidæ.

Camelus (Camel).
Auchenia (Llama).

Oreodontidæ.

*Oreodon.

Tragulidæ.

Tragulus.
Hyomoschus.

Cervidæ.

Hydropotes.
Cervulus.
Subulo (Brocket).
Cervus (Stag).
*Megaceros.
Dama (Fallow-deer).
Alces (Elk or Moose).

———

Moschus (Musk-deer).

Antilopidæ.

Dicranoceros=Antilocapra.

Bubalis.
Portax (Nylghau).
Catoblepas (Gnu).
Tragelaphus.
Oryx = Hippotragus.
Damalis.
Oreas (Eland).
Cephalophus.
Cervicapra=Kobus.
Antilope (Gazelle).
Colus=Saiga.
Calotragus.
Tetracerus.
Hemitragus=Nemorhedus.
Rupicapra.

Ovidæ.

Capra (Goat).
Ovis (Sheep).

Bovidæ.

Anoa.
Ovibos (Musk-ox).
Bos=Bubalus (Ox, &c).
Poëphagus (Yak).

PERISSODACTYLA.—Toes one or three, encased in hoofs, or, if four-toed, then the toe smaller and not touching the ground. Stomach simple ; cæcum large and sacculated.

The living species of this suborder "are widely removed from one another in many important characters ; but the intervals between them are largely filled up by an extensive series of fossil forms, commencing in the Lower Tertiary strata." A large number have been recently discovered in North America (in Utah, Colorado, Kansas, Oregon, &c.), and have been worked out principally by Marsh and Leidy.

In regard to the extinct forms of Equidæ, *Eohippus* and *Orohippus* were Eocene, *Miohippus* and *Anchitherium* Miocene, and *Hipparion*, *Protohippus*, and *Pliohippus* Pliocene. Dinocerata (= Uintatheriidæ) and Pantodontia (*Bathmodon*, &c.) constitute the Amblypoda of Cope.

Of the Rhinoceroses one form (*R. tichorhinus*) was once common in Europe.

Rhinocerotidæ.

Rhinoceros.
*Acerotherium.

Tapiridæ.

Tapirus (Tapir).

Palæotheriidæ.

*Palæotherium.

———

*Macrauchenia.

Coryphodontidæ.

*Coryphodon.
*Lophiodon.

Equidæ.

*Eohippus.
*Orohippus.
*Miohippus.
*Anchitherium.
*Hipparion.
*Protohippus.
*Pliohippus.
Equus (Horse, Ass, &c.).

Brontotheriidæ.

*Brontotherium.

Tillotheriidæ (= *Tillodontia*).

*Tillotherium.
*Stylinodon.

Uintatheriidæ (= *Dinocerata*).

*Uintatherium.
*Dinoceras.
*Tinoceras = *Eobasileus.

———

*Bathmodon.

The systematic position of the following group is very doubtful. It comprises certain very large extinct mammals from the Pleistocene deposits of South America, having affinities with the Rodentia, Bruta, and Ungulata. It forms the order Toxodontia of Owen.

TOXODONTIA.

* Nesodon.
* Toxodon.

Order X. CETACEA.

NATANTIA. MUTILATA.

Fish-like ; nostrils (spiracles) on the top of the head. Fore limbs only, fin-like and without nails. No sacrum. Pelvis rudimentary. A horizontal tail-fin without bony rays.

The Cetacea differ from all other mammals in having the nostrils, or blowholes, placed on the top of the head, and, excepting Sirenia, in having a horizontal tail-fin, not supported by bony rays. A few species have a dorsal fin. By some they are supposed to be modified Ungulata, by others modified Carnivora (Pinnipedia).

There are no external ears. The body is sparsely hairy, or it may be without hairs in the adult. There are no clavicles, and sometimes there are no teeth except in the fœtus. The dolphins have from 100 to 200 teeth. In the male narwhal a tusk is developed in the left premaxillary bone often attaining the length of ten feet ; on the right and in the female the germ-cavities are gradually enclosed by the forward growth of the bone. The two mammæ are inguinal. The placenta is diffused. To the palate are attached certain plates, or baleen ; as many as 302 rows on either side have been counted in the "right" whale [*Eschricht*]. From these are derived the whalebone of commerce. Their use is to retain the small animals on which the creature lives, until, having collected a sufficient quantity, they are swallowed, the water either making its escape at the sides or being ejected through the blowhole. The stomach is complex, divided into three or four cavities in all true Cetacea. Two little detached bones, placed near the anus, are the only vestiges of the posterior extremities.

Spermaceti is a fatty secretion found in cells in the upper part of the skull of the cachalot, as well as in the blubber. Ambergris is an odorous concretion found in the intestines.

Among the members of this order are the grampus (*Phocæna orca*), porpoise (*Phocæna communis*), dolphin (*Delphinus delphis*), [the dolphin of sailors is a tropical fish, *Coryphæna hippuris*], dolphin of the Ganges (*Platanista gangetica*), dolphin of the Amazon (*Inia amazonica*), grampus (*Orca gladiator*), the ca'ing whale or bottlenose (*Globiocephalus globiceps*), the finner, rorqual, or fin-whale (*Balænoptera sibbaldii* and *B. rostrata*), hump-backed whale (*Megaptera boops*), cachalot or spermaceti whale (*Physeter macrocephalus*), right or Greenland whale (*Balæna mysticetus*), narwhal (*Monodon monoceros*), white whale (*Beluga leucas*).

There are upwards of sixty genera for less than that number of well-ascertained species; Sowerby's whale has been placed in not less than thirteen of them. Zeuglodontidæ are found in the Eocene and Miocene deposits of North America; they differ from all other Cetaceans in having molars with two fangs. Some of the Cetacea attained a length of 70 feet.

Balænopteridæ (Fin-whales, or Rorquals).	*Physeteridæ.* Physeter = Catodon (Cachalot, or Sperm-whale).	*Delphinidæ.* Platanista. Inia.
Megaptera. Physalus. Balænoptera.	*Monodontidæ.* Monodon (Narwhal).	Delphinus (Dolphin). Orca (Grampus). Phocæna (Porpoise). Globicephalus (Bottlenose).
Balænidæ. Balæna (Whale, or Whalebone Whale).	*Hyperoodontidæ.* Ziphius. Hyperoodon.	Beluga (white whale, or white-fish).

**Zeuglodontidæ.*
*Zeuglodon.

Order XI. SIRENIA.

Body fish-like. Nostrils on the muzzle. Fore limbs only fin-like. No sacrum. Pelvis rudimentary. A horizontal tail-fin without bony rays.

The Sirenia differ from the whales in having two kinds of teeth, incisors and molars; in the position of the nostrils on the snout; in the fleshy lips, provided with short thick scattered bristles; and in the pectoral mammæ. They have also a third eyelid and salivary glands. Their limb-bones are solid; they have no clavicles, and no external ear. Their pelvic bones in their highest development "retain the size and shape of the small contiguous costal arches." The stomach is large, and the cæcum is of moderate size. The diaphragm is exceedingly oblique.

In the dugongs the males are furnished with two large incisors in the upper jaws; in the female they are arrested in their growth without cutting through the gum, and they remain through life concealed in the premaxillaries.

The Manatidæ (the only family) are herbivorous; one of its species (*Manatus americanus*) lives almost exclusively on *Pistia stratiotes*; another species (*Rhytina stelleri*), of large size, is now extinct. Other members of the order are the dugongs (*Halicore australis* and *H. indica*) and two or three species of manatee or sea-cow (*Manatus*). *Halitherium* and *Prorastomus* are from the Tertiary deposits; the latter had canine teeth in both jaws.

In the recent works of Claus and Schmarda this group is not separated from the Cetacea. Cope's Homodonta include these orders and the Edentata.

Manatidæ.

Halicore (Dugong).
Manatus.

Rhytina.
*Halitherium.
*Prorastomus.

Order XII. PINNIPEDIA.

Limbs fin-like; the posterior horizontal, directed backwards. Three kinds of teeth. Mammæ two or four, ventral.

The toes are entirely enveloped in the integument, and almost the only indication of their presence is their strong claws. The hind limbs, from their horizontal position and their close connexion with the tail, form the principal organ of progression in the water. On the land they move by a shuffling on of the body or by short leaps.

The teeth vary in number according to age; the incisors are nearly all deciduous after a time. The lips are fleshy, furnished with long bristles, and the nostrils are capable of being closed by a peculiar sphincter muscle. The brain is large. There are no clavicles.

Owing to the peculiar structure of the vena cava, the blood finds its way back to the lungs so slowly as to check the respiration, and so enables the animal to remain a considerable time under water.

The male of the sea-elephant (*Cystophora proboscidea*) has a muscular sac at the tip of the nose, which it can inflate at pleasure; a somewhat analogous appendage is placed on the head of *Stemmatops cristata.*

The common seal (*Phoca vitulina*) and the grey seal (*Halichœrus gryphus*) are the only species that can be said to inhabit these islands. Others have been mentioned, but they are either stragglers or doubtful; among them are the harp-seal (*Phoca grœnlandica*) and the walrus (*Trichechus rosmarus*).

Pinnipedia are by some regarded as a suborder of Carnivora; Claus and Schmarda rank it as an order. Two very distinct families are included. The eared seals (*Otariæ*) are by some separated from the ordinary seals. The genera, so-called, have been excessively multiplied.

Phocidæ (Seals).	Halichœrus.	Otaria = Arctoce-
Cystophora (Sea-elephant).	Pelagius = Steno-rhynchus.	phalus.
Stemmatops=Stemmatopus.	Phoca = Callocephalus.	*Trichechidæ.*
		Trichechus (Walrus).

Order XIII. CARNIVORA.

FERÆ.

Three kinds of teeth, the canines large and projecting. Toes with long sharp claws. Mammæ abdominal.

The toes are webbed in the otters; in the Felidæ the claws are retractile. Some species walk with the whole of the foot to the ground [plantigrade], others entirely on their toes [digitigrade], but some of the Viverridæ do both. A remarkable peculiarity of this order is that the lobes of the cerebrum are separated from the cerebellum by a long process arising either from the occipital or parietal bone, or from both. The clavicle is either wanting or rudimentary, but is never attached to the sternum or to the scapula. The stomach is simple, the intestines short, and the cæcum small or wanting.

To this order belong the lion (*Felis leo*), jaguar (*Felis onca*), puma (*Felis concolor*), tiger (*Felis tigris*), leopard (*Felis leopardus*), wild cat (*Felis catus*) [the origin of the domestic cat is unknown], cheetah (*Felis jubatus*), lynx (*Lynx cervaria*), hyæna (*Hyæna striata* and *H. crocuta* are the two common species), fox (*Vulpes vulgaris*), wolf (*Canis lupus*) [the dog is probably a domesticated variety], jackal (*Canis aureus*), ichneumon or Pharaoh's rat (*Herpestes ichneumon*), civet-cat (*Viverra civetta*), martin (*Mustela martes*), sable (*Mustela zibellina*), polecat (*Putorius fœtidus*) [the ferret is a domesticated variety], weasel (*Putorius vulgaris*) stoat (*Putorius ermineus*) [in its winter dress it is the ermine], mink (*Putorius lutreola*), glutton or wolverene (*Gulo borcalis*) [the American glutton is not distinct], otter (*Lutra vulgaris*), sea-otter (*Enhydris marina*), badger (*Meles taxus*), skunk (*Mephitis americana*), racoon (*Procyon lotor*), bear (*Ursus arctos*), polar bear (*Ursus maritimus*), grizzly bear (*Ursus ferox*), and sun-bear (*Helarctos malayanus*).

Prof. Flower has founded a classification of this order on the

S

characters afforded by the auditory bulla and surrounding part at the base of the skull. He divides it into three groups—Æluroidea, Cynoidea, and Arctoidea. They are given below as well as the families belonging to them. It is unfortunate that the two names Æluroidea and Æluridæ should clash, as not belonging to the same " sections."

The Pinnipedia are sometimes placed in this order. The first four families in the following list are plantigrade; the remainder, with some exceptions, are digitigrade.

ARCTOIDEA.

Ursidæ.

Prochilus.
Helarctos (Sun-
 bear).
Ursus (Bear).

Procyonidæ.

Procyon (Racoon).
Bassaris.
Nasua.

———

Cercoleptes.

Æluridæ.

Ælurus.

Mustelidæ.

Gulo.
Galictis.

Putorius (Polecat
 &c.).
Mustela (Martin).
Mellivora.
Mephitis (Skunk).
Mydaus.
Meles (Badger).
Pteronura.
Enhydra.
Lutra (Otter).

CYNOIDEA.

Canidæ.

Megalotis.
Vulpes (Fox).
Canis (Dog).

ÆLUROIDEA.

Felidæ.

Lynx (Lynx).

Felis (Cat, Lion, &c.).
*Machærodus.

Hyænidæ.

Proteles.
Hyæna.

Viverridæ.

Herpestes (Man-
 gouste).
Rhyzæna.
Crossarchus.
Arctictis=Ictides.
Prionodon.
Viverra.
Paradoxurus.
Cynogale.
Cryptoprocta.

Order XIV. QUADRUMANA.

PRIMATES. POLLICATA.

The hallux, and in many the pollex, opposable to the digits. Teeth uneven and interrupted; never more than four incisors in each jaw.

It is only a certain number that are "four-handed;" in many the anterior extremities have no thumbs, and in *Galeopithecus* there is no opposable thumb on either extremity. The canine teeth pass beyond the line of the other teeth; and in the upper jaw there is an interval in which the lower canine is received, as in the Carnivora; the upper canine passes towards or outside the lower jaw, and is sometimes a formidable tusk. Clavicles are always present.

In a natural state the Quadrumana are quadrupedal; their narrow pelvis and inability to place the sole of the hind feet to the ground, owing to the oblique articulation of the foot on the leg, disable them from walking erect except with difficulty.

The younger animals of the anthropoid species most nearly approach man in the form of the skull; as they get older they become more bestial, and the brain is smaller in proportion. The two mammæ are pectoral, except in some of the lemurs, in which they are sometimes ventral.

This order comprises two suborders—Simiæ and Prosimiæ. The latter has been separated as an order by some authors. Carus even places it between the Rodentia and the Carnivora. The Simiæ have a discoidal deciduate placenta; a face mostly naked or without hair, and flat nails on all the fingers, although there are some exceptions. The Prosimiæ have an indeciduate placenta, a hairy face, and nails mostly unguiculate. Their brain is much less convoluted (in some quite smooth) than the brain of the Simiæ; and the incisors, always four in each jaw in the Simiæ, are sometimes only two in the Prosimiæ.

In the Simiæ of the New World the nostrils are lateral and widely apart (Platyrrhini), and the pollex is not opposable to the fingers; Cebidæ have prehensile tails. In the Simiæ of the Old World the nostrils are oblique and close together (Catarrhini), and the pollex is opposable to the fingers. In the Prosimiæ the nostrils are curved or twisted (Strepsirrhini).

There are many fossil forms supposed to be allied to Lemur, especially several recently discovered in the Eocene of North America. The first remains of the higher Quadrumana appear in the Miocene.

In this order we find the gorilla (*Troglodytes gorilla*), chimpanzee (*Troglodytes niger*), ourang-outang (*Simia satyrus*), ape [Barbary] (*Inuus sylvanus*), baboon (*Cynocephalus papio*), mandrill (*Cynocephalus mormon*), spider-monkey (*Ateles paniscus*), green monkey (*Cercopithecus sabæus*), howler (*Mycetes seniculus*), and marmoset (*Hapale jacchus*). The sacred monkey of India is *Semnopithecus entellus*. Among the Lemurs are the macaco (*Lemur catta*), potto (*Perodicticus potto*), and the aye-aye (*Chiromys madagascariensis*).

The last two genera of the following list comprise the "Anthropoid apes."

PROSIMIÆ.

(Pedimana=Le-
muroidea=Strep-
sirrhini.)

Chiromyidæ.
Chiromys.

Tarsiidæ.
Tarsius.

Lemuridæ.
Nycticebus.
Galago=Otolicnus.
Stenops.
Perodicticus.
Lichanotis.
Propithecus.
Chirogaleus,
Lemur=Prosimia.

SIMIÆ.

" Platyrrhini."
Hapalidæ (Marmo-
sets).
Hapale=Iacchus.
Midas.

Pitheciidæ.
Callithrix.
Chrysothrix.
Nyctipithecus
(Night-ape).
Pithecia (Saki).

Cebidæ (Monkeys).
Cebus,
Ateles.
Mycetes.
Lagothrix.

" Catarrhini."

Cercopithecidæ.
Cynocephalus=
Papio (Baboon).
Inuus (Ape).
Rhesus.
Cercopithecus.
Colobus.
Semnopithecus.
Nasalis.
Presbytis.
Macacus.
Hylobates (Gibbon).
*Dryopithecus.

Simiidæ.
Simia=Pithecus
(Ourang-outang).
Troglodytes (Gorilla,
Chimpanzee).

Order XV. BIMANA. (Man.)

ERECTI. ANTHROPIDÆ. HOMINIDÆ. ARCHENCEPHALA.

Hallux not opposable to the digits; all the nails broad and flat. Teeth even, contiguous. Skin not covered with hair. Walks erect.

It would perhaps be more in accordance with modern views to place man as a "family" in the preceding order, seeing that he is morphologically nearer to the higher apes than the higher apes are to the lemurs. Yet, when it is considered that he alone is endowed with the power of improvable reason and of articulate speech, and that "he is the only earthly being of practically un-limited power," it may be a question whether such differences do not warrant giving to him a higher taxonomic rank. Claus, indeed, places him at the end of his work as "Der Mensch," without consigning him to any order or family, like the rest of the animal kingdom; and Schmarda omits him altogether. But no one can fail to draw "a very sharp line between man and the ape," and few "are disposed to underrate the enormous gap which separates man from the brutes."

The characters peculiar to man are numerous, but mostly adaptive. Such are the tenuity of the derm or skin; the rudimentary hairs, except on particular parts; the comparatively small face, even in the lowest savages, compared to the large size of the skull; and the even teeth without a break in the series. He is the only "plantigrade biped" known, and the only animal whose chief locomotive power is thrown on the innermost side of the foot.

That man was contemporary with the mammoth, the woolly rhinoceros, and the cave-bear can now admit of no doubt. His bones, preserved from decay by the constant dripping of water charged with carbonate of lime, have been found in many calcareous caverns in company with those of these and other extinct animals. The Engis cavern, near Liége, has yielded a skull which, being restored, is one of the most perfect that has yet been found; another, from the Neanderthal, is said to be the "most brutal of all known skulls." Yet these skulls do not differ essentially from one another or from modern types more than those of now existing races differ from each other. The Neanderthal skull stands, indeed, "in capacity very nearly on a level with the mean of the two human extremes, and very far above the pithecoid maximum" (*Huxley*).

Whatever may be said as to the unity of the species, or of the "endless diversity of opinion" that exists as to races, it is admitted that there is only one genus—*Homo*. Linnæus, in his 'Fauna Suecica' (1761), puts it at the head of his order "Magnates" [afterwards changed to Primates], under the specific name of "Homo *sapiens*," with the character "Naturæ regnorum Tyrannus."

GLOSSARY.

Abdomen. The cavity containing the intestines.

Abiogenesis. The production of living organisms without pre-existing germs.

Acephalous. Having no head.

Acetabula. The suckers on the arms of a cuttle-fish.

Acetabulum. The cavity which receives the head of the thigh-bone.

Acœlous. Without an intestinal cavity.

Acontia. See *Craspeda.*

Acrocyst. "An external sac which in certain hydroids is formed upon the summit of the gonangium, where it constitutes a receptacle in which the ova pass through some of the earlier stages of their development." (*Allman.*)

Acrodont. The attachment of a tooth by its base to the edge of the jaw.

Acromyodic. When certain muscles in birds are attached to the end of the bronchial semirings of the syrinx.

Actinomeres. The lobes lying between the ctenophores of an actinosoma.

Actinosoma. The body, simple or compound, of an Actinozoon.

Actinula. "The locomotive polypoid embryo into which, in certain genera (of Hydroida), the egg becomes directly developed." (*Allman.*)

Aculeus. The sting of bees, wasps, &c.

Adaptation. The variation which tends to fit an organ for the part it has to perform, or to enable it to meet new conditions.

Adelocodonic. The condition of a gonophore when no developed umbrella is present.

Ætiology. The study of physical causes in the origin and development of organized beings. The "doctrine of efficient as opposed to final causes."

Agamic. Non-sexual reproduction.

Agamogenesis. Discontinuous development, as when the ova is not brought into contact with the spermatozoa.

Air-bladder. See *Swim-bladder.*

Allæogenesis. A term used by Häckel to denote a form of production in the Geryoniidæ. (Now explained in another way.)

Allantois. A fœtal membrane, disappearing in the Mammalia at an early period of fœtal life, or else it is " placentiferous." In function it is respiratory.

Alternation of generations. First used by Steenstrup to designate the phenomenon of an animal bringing forth a progeny not resembling itself, but to whose descendants the resemblance returns in the second, third, or fourth generation. One is an act of reproduction, the other of development. A "successive series of individuals" which "seem to represent two species alternately reproduced" (*Owen*). "An alternation of asexual with sexual generation, in which the products of the one process differ from those of the other" (*Huxley*). "An intercalation of a proper sexual reproduction" is necessary in a true alternation, according to Dr. Allman. See also *Metagenesis.*

Alulæ. The small membranous appendages at the base of the wings posteriorly in the Diptera.

Alveoli. The sockets of the teeth in the Mammalia. In the Radiolaria they are certain vacant spaces in the sarcode, placed either within or without the capsule.

Ambulacra. The perforated spaces for the emission of the tube-feet (pedicelli) in the Echinodermata. The tube-feet themselves are sometimes called ambulacra, and that part of the plate from which they issue ambulacral spaces.

Ametabolous. Not undergoing any change.

Amnion. A fœtal membrane enveloping the embryo. It is found only in reptiles, birds, and mammals.

Amphiblastula. A stage in the development of sponges before they become fixed.

Amphicœlous. Said of vertebræ which are concave at both ends.

Amphidiscs. "Two-toothed disks, like cogged wheels, united by an axis" (*Huxley*), forming a siliceous spiculum found in certain sponge-corpuscules. "Spicula which surround the gemmula of *Spongilla*" (*Nicholson*).

Amphigonous. When qualities or characters are transmitted from *both* parents.

Analogy. A similarity of functions without a similarity of parts.

Anapophyses. Processes of the lumbar vertebræ.

Anchylosis. The union of two bones to form one bone.

Androgeny = hermaphroditism.

Androphores. The gonophores carrying the male elements in the Hydroida.

Anenterous. Without intestines.

Antennæ. Two or four movable jointed organs situated before or between the eyes in the Arthropoda. Also in some worms, but unjointed.

Antennulæ. The smaller antennæ when four are present.

Anthogenesis. "That mode of reproduction in which there intervenes a form furnishing male and female pupæ from which the sexual individuals issue." It occurs in some of the Phytophthiria.

Antigeny. Sexual dimorphism.

Antimeres. "Equivalent parts or homotypes" (*Gegenbauer*). The parts formed by the segmentation of the embryo. "They vary in number; each segment of a bilaterally symmetrical animal (Vertebrate or Arthropod) has two:" there are as many as eight in the Ctenophora. "The increase of a function may be provided for by a multiplication of organs;" "hence the next step in complexity is the formation of a chain of similar groups in succession." To each element in this chain is given the name of *metamere.* Thus "a vertebrate animal is made up of a series of successive externally-unjointed metameres, each consisting of two symmetrical antimeres" (*Macalister*, An .Morph.).

Antlia. The suctorial mouth-organ of the Lepidoptera.

Anus. The termination of the intestine.

Aorta. The artery arising from the left ventricle of the heart.

Apodemata. Certain processes in the interior of the thorax of the higher Crustacea, serving for the attachment of muscles.

Aponeurosis. The expanded tendon of a muscle.

Apophysis. A process of a bone ("a mischievous word," *Parker*).

Aproctous. Without an anal opening.

Apteria. The naked parts in the skin of birds where feathers do not occur.

Aptychi. "Plates of a shelly substance" found associated with ammonites, or sometimes lodged within the shell.

Arachnidium. The spinning-apparatus of the spiders.

Archæostomatous. When the mouth of the gastrula is retained.

Archebiosis = spontaneous generation.

Archetype. The simple primary form.

Archigony. The primitive generation of organic from inorganic matter.

Arrhenotocous. Applied to those cases in which males only are developed from the eggs. The females are partheno-genetic.

Arthrium. The minute penultimate tarsal joint of many Coleoptera.

Asteriscus. One of the otoliths in fishes.

Astomatous. Without a mouth.

Astragalus. A tarsal bone articulating above with the tibia.

Atavism. See *Reversion.*

Atlas. The first vertebra of the neck.

Atrium. The cloacal cavity of the Tunicata.

Auricle. A cavity of the heart.

Autogenous. When parts of a bone are developed from independent centres of ossification.

Autogeny. The origin of an organism from " an inorganic formative fluid."

Autophagous. When newly-born animals are at once capable of feeding themselves.

Avicularia. Organs of prehension in the Polyzoa, consisting of a movable portion, or mandible, and a corresponding fixed portion.

Axis. The second vertebra of the neck.

Azygous. Single; without a fellow.

Barbs. The small branches forming collectively the web or vane of a feather.

Barbule. The small processes on each side of the barb.

Basipodite. The small conical joint (the second) attached to the first joint of the leg of a Crustacean.

Biogenesis. The production of living from pre-existing organisms.

Biology. The science of living beings, including Zoology and Botany.

Bioplasm. Dr. Beale's name for Protoplasm.

Bioplast. See *Plastide.*

Biotome (Cobbold). "A successive life-epoch in the development of some of the lower animals," *e. g.* Entozoa.

Blastema. A mass of formative matter.

Blastocheme. "A medusiform planoblast which gives origin to the generative elements, not directly, but through the medium of special sexual buds which are developed from it" (*Allman*).

Blastocœle. The cavity of the morula.

Blastoderm. The outer membrane of the embryo.

Blastomeres. *Division-masses* or *germ-masses.* The divisions of the germ; these become cells and give rise to tissues.

Blastopore (Ray Lankester). The orifice of the invagination in certain invertebrates, either becoming a mouth or eventually closing up.

Blastosphere. Blastomeres arranged in a hollow sphere.

Blastostyle or *gonoblastidium.* "A columniform zooid destined to give origin to generative buds" (*Allman*).

Brachycephalous. When the breadth of the head is more than its length.

Branchiæ=gills. The organs in which the venous blood is oxygenized.

Bronchi. The branches of the windpipe conveying the air to the lungs.

Bulbus arteriosus. The dilated base of the arch of the aorta.

Byssus. The filamentous substance secreted by the mouth of certain bivalve Mollusca.

Cæcum. A blind pouch opening into the duodenum.

Cainozoic. The Tertiary period.

Calamistrum. Two rows of movable spines on the metatarsal joints of each posterior leg of certain spiders form an apparatus called a calamistrum.

Calcaneum. The heel-bone or os calcis.

Calicle. The cup-like excavation terminating the theca of a corallite.

Calycle or *hydrotheca.* The receptacle in which the polypites are lodged in the Calyptoblastea.

Canthus. The angles or corners of the eyes. In certain insects it is a process of the clypeus partially or completely dividing the eye.

Carapace. The dorsal plate of the Crustacea and Chelonia.

Carpodite. The fifth joint of the leg of a Crustacean.

Cell. (1) In its original condition " a naked lump of protoplasm with an imbedded nucleus," and with or without an external membrane. (2) The interneural spaces in the wings of insects.

Cellulose. A peculiar substance forming the cell-wall of plants ; it is found also in the tests of the Ascidioida.

Cænogenesis. Embryonic adaptation.

Central capsule. A porous membrane separating the sarcode of the Radiolarians from the " yellow cells."

Centrum. The body or common centre of a vertebra.

Ceratode. The horny substance of sponges.

Cercariæ. The tadpole-like larvæ of the Trematode worms and of many Ascidians.

Cerci. Setaceous or filiform appendages attached to the last segment of the abdomen in certain Orthoptera.

Cere. The naked skin at the base of the bill in some birds.

Cerebellum. The posterior portion of the brain.

Chelæ. The "claws" or anterior pair of thoracic legs of the Crustaceans, generally of large size and furnished with two "fingers," only one of which is movable.

Chelicera. The prehensile "pincer-ended" claws placed on each side of the mouth of scorpions. They are supposed to be modified antennæ.

Chevron-bones. Small bones placed below and between the caudal vertebræ and protecting the artery.

Chiasma. The expansion formed by the union of the optic nerves.

Chitine. The substance composing the exoskeleton of insects.

Chlorophyll. The green colouring-matter of leaves, found also in the Infusoria, Turbellaria, &c.

Choanæ. The nasal cavities.

Chondrine=gristle.

Chorda dorsalis. See *Notochord.*

Chorion. The outer membrane of the ovum.

Chorology. The study of the local distribution of animals over the earth.

Chromatophores. Minute sacs containing pigmentary matter.

Chyle. A milky fluid, the nutrient portion of the food.

Chyme. The digested food as it passes from the stomach.

Cicatricula or "tread," a peculiar opaque spot on the embryo.

Cicatrix. The truncated portion of the apex of the basal joint of the antennæ in some of the Longicorn Coleoptera.

Cilia. Minute hair-like bodies, which in the lower forms of Invertebrata are organs of locomotion.

Cinclides (sing. *Cinclis*). Apertures in the walls of the somatic cavity of the Actiniæ for the emission of craspeda and acontia.

Cirri. Curled appendages on the feet, mouth, &c. in many animals.

Clavicle. The collar-bone.

Clavus. The basal inner portion of the hemelytron in the Hemiptera.

Cloaca. The common efferent opening in birds and many other animals.

Clypeus. The part, often very indistinctly marked off, to which the upper lip and its membrane is attached in most mandibulate insects ; it is often called the epistome.

Cnidæ. See *Trichocysts.*

Coarctate. Applied to an insect pupa where it gives no indication of the parts it covers.

Coccoliths. Minute calcareous concretions formed at the end of the contractile processes of certain Radiolaria ; when they are massed together they are called coccospheres.

Coccyx. The anchylosed terminal tail-bones in birds and some mammals.

Cocoon. The outer covering, whether of silk or other material, of the pupæ of certain insects.

Codonostoma. "The orifice of the umbrella in the Medusæ, through which its cavity communicates with the external water" (*Allman*).

Cœloma. The general body-cavity.

Cœnenchyma. The common calcareous tissue which connects the sclerodermic coralla of certain Actinozoa.

Cœnœcium. See *Polypary.*

Cœnosarc. "The common flesh or trunk which unites and binds together the polypites in a compound zoophyte" (*Hincks*).

Colon. The large intestine opening into the rectum.

Columella. The axis of a spiral univalve. The centre of the thecæ in a corallite.

Commensal. An animal that lives with but does not feed on its host.

Concha. The external ear.

Condyle. The articulating surface of a bone.

Conjugation. "The coalescence of two similar masses of protoplasm;" not supposed to be of a sexual nature.

Conjunctiva. The mucous membrane of the eye, covering the anterior surface and reflected internally on the eyelids.

Connexivum. The lateral more or less expanded border of the abdominal segments in certain Heteropterous Hemiptera.

Coracoid. A process of the scapula or sometimes a separate bone.

Corallite. The corallum secreted within the body of a polype. There may be a single corallite, or several connected by a cœnenchyma.

Corallum. The skeleton or hard structure deposited in the tissues or by the tissues of the coralligenous Actinozoa.

Corbel. A hollow or cavity partially closed by a plate in which the tarsus is inserted in certain Coleoptera.

Corbiculum. The dilated posterior tibia of the Apidæ is sometimes so named.

Corbulæ. Basket-shaped receptacles which enclose the gonangia in *Aglaophenia*.

Corium. The middle portion of the hemelytron in the Hemiptera, between the clavus and the cuneus.

Corneule. Applied to the transparent "segments" which defend the eyes of insects.

Corpus callosum. A layer of transverse fibres forming the great commissure of the brain connecting the two hemispheres.

Correlation. The mutual relation or association of phenomena or of parts. The correlation of parts does not always imply correlation of development.

Cotyledonary. When the villi of the placenta are collected into bundles.

Cotyloid cavity. The opening in which the coxa of insects is placed.

Coxa. The basal joint of the leg in insects.

Coxopodite. The basal joint of the leg of a Crustacean.

Craspeda. Convoluted cords formed in the Actiniæ and furnished with thread-cells.

Crepitaculum. A talc-like spot at the base of the upper wings in certain Locustidæ.

Ctenocyst. A peculiar body in the Ctenophora.

Ctenoid. Applied to fish-scales with a toothed or spinous hinder margin.

Ctenophores. Meridional bands, eight in number, bearing comb-like fringes ; the organs of locomotion in the Ctenophora.

Cuneus. A portion of the hemelytron of certain Hemiptera between the corium and the posterior membranous portion.

Cycloid. Applied to fish-scales with a rounded entire hinder margin.

Cytoblast. See *Nucleus.*

Cytode. A plastide without a nucleus. A plastide with a nucleus is a cell. Prof. Huxley holds that "the primary form of every animal is a nucleated protoplasmic body, *cytode* or *cell.*"

Cytogenous. Producing cells.

Cytostome. The point where the ingestion of food takes place in the flagellate Infusoria.

Dactylopodite. The seventh or terminal joint, exclusive of the " fingers," in the leg of a Crustacean.

Decidua. " The modified mucous membrane of the pregnant uterus."

Degradation. Rudimentary or abortive structural development. Often due to parasitism. Not to be confounded with arrest of development.

Delamination. "The splitting into two layers of cells of a primitively single-layered blastoderm" (*Huxley*).

Dentine. The tissue forming the body of the tooth.

Derivative theory as opposed to Natural Selection, holds "that every species changes, in time, by virtue of inherent tendencies thereto."

Dermatosis. When the derm or skin forms a bony plate.

Dertrum. The apex of a bird's bill.

Deuterostomatous. When the mouth of the gastrula is secondary.

Deuterozooid or *proglottis.* Zooids produced by gemmation from zooids.

Dialysis. The separation of parts previously joined together.

Diaphragm. The muscle separating the thorax from the abdomen.

Diaphysis. Ossification proceeding from the centre of a long bone.

Diapophysis. The upper articular transverse costal process of certain vertebræ.

Diastema. An interval in the line of teeth.

Digitigrade. When an animal walks on its toes.

Dimorphous. A species having two forms not depending on sex.

Diœcious. The sexes in separate individuals.

Diphycercal. The tail in fishes being equal above and below, with the vertebral axis in the centre.

Diphyodont. When the earlier teeth are replaced by a second . set.

Diphyzooid. A reproductive group of organs detached from the cœnosarc of certain Calycophoridæ.

Diploë. The cancellous layer between the two plates of a flat bone.

Diverticulum. A blind tube springing from the side of another tube.

Dolichocephalous. When the length of the head is more than its breadth.

Dorsigrade. When a mammal walks on the back of its toes, as in certain armadillos.

Dualistic theory holds that creation was definite and purposive.

Duodenum. The first portion of the small intestine.

Dysteleology. The study relating to the "purposelessness" of structure or of organs.

Ecderon. The "external plane of growth" of the ectoderm of the Actinozoa.

Ecdysis. Shedding the skin or moulting.

Echinopædium. Prof. Huxley's name for the "worm larva" of the Echinodermata.

Ecthoræum. A thread-like body continued and capable of being discharged from the cnidæ of the Actinozoa.

Ectocyst. The outer layer of the cœnœcium of the Polyzoa.

Ectoderm (epiblast of the embryo). A multicellular membrane, " the result of the segmentation of the vitellus in a true ovum " (*Allman*). The external tegumentary layer of the Metazoa.

Ectopterygoid. One of the lateral palatine bones in certain reptiles. It is peculiarly developed in the Crocodilia.

Ectosarc. The outer layer of sarcode in the Protozoa.

Ectostosis. Ossification proceeding from without to within.

Elytra. The upper or anterior wing-cases of the Coleoptera. The term is also applied to the scales on the back of certain Annelida.

Embolium. A part of the corium in the hemelytra of certain Hemiptera.

Emboly. "Invagination," the formation of a cavity in the embryo.

Embryo. The animal in the egg or in the womb; but it is also sometimes applied to the young larva. " We look at the embryo as a picture, more or less obscured, of the progenitor, either in its adult or larval states, of all the members of the same great class."

Embryology. The study of the embryo.

Empodium. That part of the last tarsal joint in insects to which the claws are attached.

Enamel (Encaustum). The hardest constituent, when it exists, of the tooth.

Endoderm (*hypoblast* of the embryo). The inner tegumentary layer of the Metazoa.

Endoplast. The probable analogue, according to Huxley, in the Protozoa of the nucleus of the Metazoa.

Endopleurite. That part of the apodema of a Crustacean which arises from the interepimeral membrane.

Endopodite. An inner filamentous appendage attached to the basal joint of some of the Crustacea.

Endosarc. The inner layer of sarcode in the Protozoa.

Endoskeleton. The internal hard or bony structure.

Endosternite. That part of the apodema of a Crustacean that arises from the intersternal membrane.

Endosteum. The vascular tissue lining the medullary cavity of the long bones.

Endostoma. A part behind the labrum in the Crustacea.

Endostyle. The longitudinal fold in the pharynx of Ascidians.

Enterocœle. The "perivisceral cavity" of the Echinoderms, Mollusca, &c.

Environment. "The totality of all surrounding agencies and influences" (*Mivart*). "A term of the most comprehensive kind, embodying, in every case that it is used, an assemblage of conditions presenting an amount of complexity that is not only inconceivable but wholly unnameable " (*Romanes*).

Eocene. The earliest Tertiary epoch.

Ephippium. The case in which the winter eggs of the Daphniidæ are deposited.

Ephyræ. The disk-like segments which gradually fall off from the "hydra-tuba" of certain Hydrozoa, growing often to a large size, and developing organs.

Epiblast in the embryo is the ectoderm in the adult; in the higher animals the latter becomes the epidermis.

Epiboly. An occasional form of growth in the embryo (the epiblast over the hypoblast).

Epicœle. The "perivisceral cavity" of the Ascidians and Vertebrata.

Epidermis. The outer skin or cuticle of the higher animals.

Epigenesis. The doctrine that organic development depended upon the juxtaposition of molecules according to "the operation of a developmental force."

Epiglottis. A cartilaginous valve placed in front of the larynx which it closes when in the act of swallowing.

Epimera. Lateral pieces of the thorax placed behind the episterna.

Epiphragm. The calcareous secretion of the foot in snails, closing the aperture of the shell during hibernation.

Epiphyses. The separately ossified ends of a long bone.

Epipleura. The sides of the elytra in Coleoptera, they are not commonly marked off from the dorsal portion.

Epipodia. Appendages of the foot in some Mollusca. In the Pteropoda they are wing-like expansions from the head.

Epipodite. A process of the basal joint of the legs in certain Crustacea.

Episterna. Lateral pieces of the thorax, above or outside the cotyloid cavities in the Arthropoda.

Epistoma. The part, not always apparent, connecting the upper lip to the clypeus in the mandibulate insects, or the part above the mouth generally. In the Crustacea it seems to answer to the clypeus. In the Polyzoa it is a lip or valve placed over the mouth of the polypide.

Epitheca. A continuous external layer of the thecæ of the corallites in some of the Zoantharia.

Epithelium. The thin membrane that covers the mucous membranes.

T

Epizoic. Parasitic on an animal.

Ethmoid. A bone between the two orbits.

Ethnology. The science of Race. The study of the varieties of mankind.

Etiology. See *Ætiology.*

Etology. The study of the general laws that contribute to form the character of individuals and communities.

Evolution. The descent of species, "each within its own class or group, from common parents" (*Darwin*). "A change from indefinite, incoherent homogeneity, to a definite coherent heterogeneity, through successive differentiations and integrations" (*Herbert Spencer*). "Evolution teaches us that at a certain period in the history of this planet such albuminoid substances as protoplasm came, by gradual building up, into existence" (*Ray Lankester*). "The only perfectly safe foundation for the doctrine of Evolution lies in the historical, or rather archæological, evidence that particular organisms have arisen by the gradual modification of their predecessors, which is furnished by fossil remains" (*Huxley*). The difficulty of explaining the existence of the myriads of lowest and almost structureless animals has led Dr. Bastian and others to the belief that "living matter is continually coming into being."

Exocorium. The narrow portion of the hemelytron of certain Hemiptera bordering the corium externally.

Exogenous. A term applied to bones that are developed from previously ossified parts.

Exoplasm. An expansion of the so-called cuticle in certain Protozoa.

Exopodite. An inner filamentous appendage attached to the basal joint of certain Crustacea.

Exoskeleton. The external hard integument of animals.

Exothecæ represent externally the dissepiments in the thecæ of a corallite in some Zoantharia.

External influences. See *Environment.*

Fabellæ. Sesamoid bones in the gastrocnemius muscle of a dog.

Facies. The face, also the outside figure (*statura* or *habitus*); in the latter sense Huxley proposes the word "*Metope.*"

Falces. The poison-fangs of spiders.

Femur. The thigh-bone.

Fissiparous. Asexual generation by division into two parts.

Flagellum. A hair-like body differing from a cilium in its greater length.

Foramen magnum. The large opening in the occipital bone where the spinal cord joins the brain.

Forceps. A pair of movable anal appendages, as in the earwig.

Fornix. Forms part of the floor of the left ventricle of the brain.

Frenum. A fold of skin, sometimes, as in certain Cirripedia, bearing ova.

Fulcra. Spiny scales on the ns of certain Ganoid fishes.

Funicle. That part of the antennæ of certain insects between the scape and the club.

Gamogenesis. Sexual reproduction.

Ganglion. A knot (or centre) of nervous matter.

Gastræa. The unknown or hypothetical stock, according to Häckel, from which the gastrula is derived.

Gastrocnemius. The large muscle of the calf of the leg.

Gastrula. The larval form from which all the Metazoa are supposed to be descended. It is said by Agassiz to be only another name for planula; it is, however, a later development.

Gemmation, or budding, occurs when a small portion of the parent is detached and develops into the likeness of its parent.

Gemmiparous. Forming buds.

Gemmules. "Spores," or capsules containing protoplasmic cells in certain sponges. These cells develop into the mature form, and are one of the forms of asexual growth.

Geneogenesis (Quatrefages). Apparently synonymous with Parthenogenesis.

Geographical distribution of living beings is concerned with the areas of the earth's surface within which groups of different kinds of organisms exist which are not found elsewhere. For plants botanists give from twenty-five to thirty such areas or provinces; a much smaller number suffice for the zoologist. The most usually, or rather only, adopted provinces are those proposed by Mr. Sclater for birds, but which are found to be well adapted *generally* for the whole animal kingdom (Proc. Linn. Soc. ii. p. 130, 1857). He divides the earth's surface into six regions:—(1) The Palæarctic Region—Africa north of the Atlas, Europe, Asia Minor, Persia, and Asia generally north of the Himalaya range ?, Northern China, Japan, and the Aleutian Islands. Area about 14,000,000 square miles. (2) Æthiopian Region—Africa south of the Atlas range (south of·the Sahara would have been better), Madagascar, Bourbon, Mauritius, Socotra, and probably Arabia (?) up to the Persian Gulf. Area about 12,000,000 square miles. (3) Indian Region—India and Asia generally south of the Himalayas, Ceylon, Burmah, Malacca, and Southern China, Philippines, Borneo, Java, Sumatra, and adjacent islands. Area "perhaps" 4,000,000 square miles. (4) Australian Region—Papua and adjacent islands, Australia, Tasmania, and Pacific islands. Area "perhaps" 3,000,000 square miles. (5) Nearctic Region—Greenland and North America down to centre of Mexico. Area "perhaps" 6,500,000 square miles. (6) Neotropical Region—West-India Islands, Southern Mexico, Central America, and whole of S. America, Galapagos Islands, Falkland Islands. Area about 5,500,000 square miles. The weak points of this scheme appear to be the separation of Papua or New Guinea from the Indian archipelago, and the omission of New Zealand and the Pacific Islands. These, as well as Madagascar and Japan, are "satellite" provinces, having too many endemic forms to be included in the "regions" in which they are geographically situated. The Vertebrata, Prof. Huxley thinks, "are so distributed at the present day as to mark out four great areas or provinces of distribution." "These are:—1. The *Arctogæal*, including North America, Europe, Africa, and Asia as far as Wallace's line, or the boundary between the Indian and Papuan divisions of the Indian archipelago. 2. The *Austrocolumbian*, comprising all the American continent south of Mexico. 3. The *Australian*, from Wallace's line to Tasmania. 4. The *Novozelanian*, including the islands of New Zealand." The late Andrew Murray maintained that "all the Coleoptera in the world are referable to one or the other of three great stirpes." "These are:—1, the Indo-African stirps; 2, the Brazilian stirps; and 3, 'the microtypal stirps.'" The first included the Indian archipelago and New Guinea; "the Brazilian stirps inhabits South and Central America east of the Andes and north of the river Plate." In the microtypal stirps he included Europe, Asia north of the Himalayas, North America, Peru, Chili, New Zealand, Australia, &c. He held that the fauna of one class was not to be judged by the fauna of another, and that the peculiarities of geographical distribution were only to be accounted for on the supposition of "continuity of soil at some former period." (Proc. Linn. Soc., Zool. xi. 1 *et seq.*)

Germ-cell or *Germ-vesicle.* "The first nucleated cell that appears in the impregnated ovum " (*Owen*).

Germ-lamellæ. The two primary layers of the embryo—epiblast and hypoblast—in the Metazoa ; in the great majority a third layer (*mesoblast*) is developed.

Gigerium = gizzard. The muscular stomach of certain birds and insects.

Glabellum. The central ridge in the shield of the Trilobites.

Gland. An organ that secretes certain constituents of the blood, which are then voided by a duct.

Glossarium. The middle portion of the suctorial proboscis in the Diptera.

Glottis. The opening of the larynx.

Gnathites. The masticatory organs of the Crustacea.

Gnathopod. See *Maxillipedes.*

Gonangium. A "receptacle" in which, in some of the Hydrozoa, planoblasts or sporosacs are developed.

Gonoblastidium. See *Blastostyle.*

Gonocalyx. The swimming-organ of the gonophore of a Hydrozoon.

Gonocheme. A sexual medusa. "A medusiform planoblast, which gives origin directly to the generative elements" (*Allman*).

Gonophore. "The ultimate generative zooid (in the Hydrozoa), which gives origin directly to the generative elements, ova or spermatozoa" (*Allman*).

Gonosome. The assemblage of sexual zooids in the Hydrozoa.

Gonotheca. A peculiar ovigerous capsule in some of the Hydrozoa.

Gynæcomasty. Milk-secretion in the breast of man.

Gynophores. The generative buds of the Hydrozoa which contain the ova only, not the spermatozoa.

Habit. See *Facies.*

Hæmal. Connected with the blood-system.

Hæmapophyses. Processes of the vertebræ; in the Reptilia they form the abdominal ribs.

Hallux. The great toe, or the innermost digit.

Halteres. Small clavate filamentous organs, one on each side the metathorax of the Diptera ; supposed to represent the posterior wings.

Haustellum. The suctorial proboscis in the Diptera.

Hectocotylus. One of the arms of a cuttle-fish modified into a reproductive organ.

Hemelytra. The upper or anterior wings of Hemipterous insects.

Hemimetabolic. Incomplete metamorphosis.

Heredity. The tendency (mental or physical) which asserts itself in successive generations.

Heterocercal. When the upper and lower lobes of the tail of a fish are dissimilar.

Heteroplast. The dissimilarity in the cells of a group of tissues, as in muscles, nerves, &c.

Heteroplasty. "The method whereby physiological division of labour is accomplished" (*Macalister*).

Hexicoloyg. The study of the inter-relations of animals.

Hippocampus minor. An elevation in the posterior cornu of the lateral ventricle of the brain.

Histology. The minute anatomy of the tissues.

Homocercal. When both lobes of the tail of a fish are alike.

Homogeny (Ray Lankester) Similarity of structure due to descent from a common ancestor.

Homologue. Identity of an organ in different animals under every kind of form and function.

Homomorphous. Having the same form.

Homoplast. A structure which is supposed "to have grown alike in obedience to the influence of similar external causes acting on similar innate powers" (*Mivart*).

Homoplaxy (Ray Lankester). Similarity of structure due to adaptation.

Homotypes. Analogical parts.

Hydatid. A pathological product, caused by entozoic worms, consisting of a cyst containing a watery fluid.

Hydranth. See *Polypite.*

Hydra-tuba. One of the earlier forms of certain Hydrozoa developing buds, and passing into the "Scyphistoma"-stage.

Hydrocaulus. "All that portion of the hydrophyton which intervenes between the hydrorhiza and the hydranth" (*Allman*).

Hydrocysts. Peculiar sacs which, with "groups of gonophores, are borne upon a common stem, and constitute a gonoblastidium or blastostyle."

Hydrœcium. A sac attached to the nectocalyx of some of the oceanic Hydrozoa.

Hydrophyllia. The plates protecting the polypites of the oceanic Hydrozoa.

Hydrophyton. "The common case of the trophosome by which its zooids are connected into a single colony" (*Allman*).

Hydrorhiza. "The proximal end of the hydrophyton, by which the colony fixes itself to other bodies" (*Allman*).

Hydrosoma. The entire colony of the Hydrozoa.

Hydrotheca. See *Calycle.*

Hyoid. The bone of the tongue.

Hypermetamorphosis. When an insect passes through several larval stages.

Hypoblast. The inner mucous layer of cells of the blastoderm, the endoderm of the adult; in the higher animals the latter becomes the epithelium.

Ileum. The small intestine opening into the colon.

Ilium. The large pelvic bone on each side of the sacrum.

Imaginal disks. Centres of formative tissue in the larva of certain insects, especially Diptera, which give rise to the legs, wings, &c.

Imago. The perfect insect.

Incus. One of the bones of the ear.

Individual. Physiologically a single form, morphologically an entirety of independent beings, the result of the development of a single ovum.

Individuation (Mivart). The power which differentiates and assimilates "all that comes to it (the germ) into a definite and predeterminate issue."

Infundibulum. A tubular organ found in the Cephalopoda. Also one of the gastric cavities of the Ctenophora.

Ingluvies. The crop—a dilatation of the œsophagus in birds.

Ink-bag. In the Cephalopoda an oval or pyriform glandular sac, placed near the liver or within it, and secreting a dark fluid.

Inquiline or *commensal.* A tenant, not a parasite; an animal that dwells with, not at the expense of, its host.

Instinct. An inherent cause of doing not depending on memory or observation. "Inherited memory."

Intercalary (Huxley). When intermediate forms "do not represent the actual passage from one group to another."

Intercentra. Spaces between the centra of the vertebræ.

Ischiopodite. The third joint in the leg of a Crustacean.

Kainozoic. See *Cainozoic.*

Keratode. See *Ceratode.*

Labium. Lower lip of an Arthropod attached to the mentum.

Labrum. Upper lip of an Arthropod attached to the clypeus (epistome of some writers), either with or without a membranous connexion.

Larva. A rather indefinite word, generally used for all animals undergoing a metamorphosis, for the state in which they emerge from the egg.

Larynx. The upper part of the windpipe.

Lemma. The primary or outer layer of the germinal vesicle.

Life. "Organization in action" (*Beclard*). "The sum total of the functions which resist death" (*Bichat*). "The movement of the bioplasm;" "the state of action peculiar to an organized body or organism" (*W. B. Carpenter*). "A series of definite and successive changes, both of structure and composition, which take place within an individual without destroying its identity" (*G. H. Lewis*). "The definite combination of heterogeneous changes, both simultaneous and successive, in correspondence with external co-existences and sequences" (*Herbert Spencer*). The last, speaking of the coordination of actions, says, "an arrest of coordination is death, and imperfect coordination is disease." Life is generally regarded as "a mode of energy;" in the Rotifers it may be dormant for years.

Ligula. See *Labium.*

Lingua. The tongue. The term is sometimes applied to a part of the sucking-apparatus of insects, and to the "inner integument" of the labrum in some Orthoptera, &c.

Lipostomy. Absence of a mouth.

Lissotrichous or *Liotrichous.* Having straight smooth hair. The hair is cylindrical, and a section of it circular.

Lithocyst. A cavity containing mineral particles; supposed to be subservient to hearing. It occurs in the Cœlenterata.

Loculi. Certain spaces in the corallite of an Actinozoon between the vertical septa.

Lophophore. A ridge on which the tentacles are placed in the Polyzoa.

Lores. A stripe on each side between the bill and the eye in birds.

Lorica. A shield. In the Rotifers a cylindrical gelatinous shell, into which they can withdraw their bodies.

Lymphatic vessels or absorbents. Minute and delicate vessels which collect the products of digestion and detrita of nutrition and convey them into the venous circulation. The *lacteals* are the lymphatics of the small intestines.

Madreporiform tubercle or *madreporite.* A wart-like body placed externally on the aboral surface of starfishes. Its use is unknown.

Malacology. The study of the Mollusca.

Malpighian tubes. Delicate vessels opening into the intestines in most insects; they vary from four to a hundred.

Mandibles. The upper and outer pair of jaws in the Arthropoda; they correspond to the lower jaw of the Vertebrata.

Manubrium. (1) The process in the middle of the anterior border of the sternum in mammals and birds (also *Præsternum*). (2) In the Hydrozoa the central tubular body attached to the bell or umbrella, the other extremity bearing the mouth.

Marsupium. The abdominal pouch in the Marsupial. See *Pecten.*

Mastax. The muscular bulb comprising the biting- and grinding-organs of a Rotifer.

Materialism. The denial of "spiritual substances" (*Johnson*). Scientific materialism affirms "that every effect has its cause, and every cause its effect." It rejects miracles and all supernatural processes. Moral or ethical materialism is based on material enjoyment and the gratification of the senses, and it has no place among naturalists and philosophers. (See Hist. of Creat. i. 36.)

Maxillæ. The lower pair of jaws in the Arthropoda. The upper jaw only in the Vertebrata.

Maxillipedes or *foot-jaws.* The three posterior of the six pairs of appendages of the mouth of a Crustacean; the two following pairs are the maxillæ, the last the mandibles. See *Siagonopoda.*

Meconidia. Peculiar sacs of the hydroid genus *Gonothyræa.*

Medulla. The ordinary marrow of bones. *Medulla spinalis,* the spinal cord.

Medulla oblongata. The upper enlarged portion of the spinal cord where it joins the brain.

Megalopa. The latest larval stage in the development of the higher Crustaceans.

Membrana nictitans. The third eyelid of birds, a duplicature of the conjunctiva; it is found also in sharks, frogs, in some reptiles, and in many mammals; in man and monkeys it is represented by the "plica semilunaris."

Menisci. Cartilaginous rings between the vertebræ of some birds.

Mentum. The part to which the lower lip of many Arthropods is attached.

Meroblastic. When, as in birds, the germ and yolk of the egg are separate.

Meropodite. The long fourth joint in the leg of a Crustacean.

Mesenteries. Vertical partitions in the somatic cavity of the Actinozoa.

Mesentery. The membranous connection between the small intestines and the posterior wall of the abdomen.

Mesoblast. A cellular layer between the epiblast and the hypoblast; in the adult the mesoderm.

Mesoderm. The part between the ectoderm and the endoderm.

Mesomyodic. In birds, when certain muscles are attached to the middle of the bronchial semirings.

Mesothorax. The intermediate segment of the thorax of insects; the mesosternum corresponds to it beneath.

Mesozoic or Secondary period in geology; it includes the Trias, Oolitic, and Cretaceous formations.

Metabolic. Complete metamorphosis.

Metacarpus. The assemblage of bones between the wrist and the fingers.

Metagenesis. "When the produced zooid is dissimilar to the producing zooid" (*Greene*). "Changes of form which the representative of a species undergoes in passing by a series of successively generated individuals from the egg to the perfect state" (*Owen*). See also *Alternation of generations.*

Metameres. Coalesced segments which have lost their primitive distinctness. See *Antimeres.*

Metamorphosis. Changes undergone by the individual before attaining the perfect state.

Metapodium. The posterior part of the foot in the Mollusca.

Metapophyses. Lateral processes of the vertebræ.

Metasternum. The under part of the metathorax of an insect.

Metasthenic. Having the maximum power in the posterior extremities.

Metastoma. The labium so-called of a Crustacean.

Metatarsus. The assemblage of bones between the tarsus and the toes.

Metathorax. The third or posterior segment of the thorax in insects.

Microbia. Extremely minute life-producing organisms in the air.

Micropyle. An aperture in the ovum through which the male element enters.

Mimetic. When animals of different groups resemble one another.

Mimicry or *imitation.* A repetition of form and colour. In many cases it has been shown by Mr. Bates, who first called attention to this subject, that this superficial resemblance acts as a safeguard to the form most exposed to attack. Mr. Darwin thinks it is "only explicable on the theory of descent;" but this does not explain the resemblance of many insects to leaves and twigs of trees, sometimes as if covered with moss, or of others to the excreta of birds, &c.

Miocene. The middle Tertiary epoch.

Molecule. "The smallest possible portion of a particular substance" (*Clerk-Maxwell*). "Every atom is a molecule, but every molecule is not an atom."

Monadiary. The common envelope of a monad assemblage.

Monaxial. When the axis is in one direction only.

Monism. The descent of all organic beings from one primitive stock.

Monistic theory holds that creation was the product of natural forces.

Monœcious. When the male and female are associated in one organism.

Monomyary. When a bivalve shell has only a single muscle.

Monophyletic. Derived from a single form.

Monophyodont. When only one set of teeth is developed.

Monoplast. A naked cell.

Morphology. "The history of the modification of form which the same organ undergoes in the same or in different organisms" (*Owen*). "The law of form or structure independent of function" (*Dallas*).

Morphone. A morphological element.

Morula. "The multicellular blastosphere from which the gastrula is developed."

Muffle. The naked part of the nose in the cow, dog, &c.

Mutable types. (1) Those which have undergone modification of structure during geological time; they "all belong to the most differentiated members of the classes" (*Huxley*). (2) In modern times there are numerous forms which, from inherent causes, deviate from the parental type, "the whole organization," as Mr. Darwin puts it, "having a tendency to vary."

Myelon. The spinal cord.

Myonine. The material of muscle.

Myophane. A striated layer in Infusoria, supposed to represent muscle.

Myotomes. The vertical flakes of muscle in fishes. (*Myotoma = Myocomma* of Owen.)

Natural selection is the theory that the "origin of species" is due to the "preservation of favourable individual differences and varieties, and the destruction of those which are injurious," and in "the accumulation of innumerable slight variations, each good for the individual possessor" (*Darwin*). Geological research, Mr. Darwin thinks, "does not yield the infinitely many fine gradations between past and present species required on the theory; and this is the most obvious of the many objections which may be urged against it." Mr. Darwin further observes that he has "probably attributed too much to the action of natural selection or the survival of the fittest." Prof. Mivart goes further, and asserts that "natural selection utterly fails to account for the conservation and development of the minute and rudimentary beginnings, the slight and insignificant commencements of structures, however useful these structures may afterwards become." It is, perhaps, scarcely necessary to observe that the origin of species has nothing to do with the origin of life.

Nauplius. The earliest larval form of Crustaceans.

Nectocalyx. The swimming-bell of a Medusa.

Nematocysts, Trichocysts, or *Cnidæ.* Peculiar cells in the Actinozoa containing thread-like bodies having a stinging or urticating power.

Nematophores. Appendages of the cœnosarc of Plumulariidæ containing thread-cells.

Neossology. The study of the *nestlings* of birds.

Nervures. The hollow tubes supporting the wings of insects, and acting as organs of circulation and respiration.

Neural. Relating to the nervous system.

Neurapophyses. The spinous processes of the vertebræ.

Neurilemma. The membrane investing a nerve.

Neuroglia. The connective network of the eye.

Neuropodium. The ventral portion of the parapodium of an Annelid.

Nisus formativus. One of the old-fashioned phrases, "self-deceiving, world-beguiling simulacra of science."

Nomogeny. The law of coming into being, *i. e.* of creation.

Nothæum. (Qu. νῶτος or νόθος?) A name for the mantle of the Mollusca.

Notochord or *chorda dorsalis.* "A cellular rod which is developed in the embryo" of the Vertebrata, and is ultimately replaced by the spinal cord.

Notopodium. The dorsal portion of the parapodium of an Annelid.

Nucleolus. A minute particle in the middle of the nucleus. In the ovum it is the "germinal spot."

Nucleus. The central particle in the elementary cells of animal tissues. In the ovum it is the "germinal vesicle."

Nymph. The *active* pupa of certain insects.

Ocelli or *stemmata.* Simple or supplementary eyes in insects and spiders.

Odontophore. See *Radula.*

Œcoid. A name for a blood-disk.

Œdagus. The *membrum virile* of insects.

Œsophagus. The gullet, connecting the mouth to the stomach.

Omostegite. The posterior portion of the dorsal plate of a Crustacean.

Oology. The study of eggs of birds.

Ontogeny. The development of the individual from the germ-cell to the mature form.

Ontology. The study that relates to the being of an individual organism. Formerly a synonym for metaphysics.

Oœcia. Cells for receiving the ova of the Polyzoa. They are also called ovicells and ovicysts.

Operculum. (1) The gill-cover of fishes. (2) The disk closing in the mouth of most univalve shells.

Ophthalmite. The peduncle supporting the eye of the Decapod Crustacea.

Opisthocœlous. Said of vertebræ which are concave only behind.

Optic lobes or *corpora quadrigemina* (or *bigemina*). Oval or spherical bodies giving rise, wholly or in part, to the optic nerve.

Optic thalami. The inferior ganglia of the cerebrum on which the optic nerves rest.

Organ-systems are the bone-system, nerve-system, &c.

Orientation. The relative direction of parts.

Orthognathous. When the jaws do not project and the teeth ai perpendicular. The highest type of the Vertebrata.

Oscula. The large exhalant apertures of a sponge.

Ossification. The conversion of tissue into bone by the deposition of earthy matter.

Osteine. The tissue of bone.

Osteocomma or osteomere, bone-segment, sclerotome, or vertebra.

Ostioles. The smaller inhalant apertures of a sponge.

Otoconium. The ear-dust of the higher Mammalia, composed of calcareous particles.

Otocyst. A vesicle, often containing otoliths, in some of the Invertebrata, and subservient to hearing.

Otoliths. (1) The internal ear-bones of fishes. (2) Calcareous bodies connected with hearing in many of the lower animals.

Ovary. The organ in which the ova are produced.

Ovicell, Ovicyst. See *Oœcia.*

Ovipositor. A tubular organ possessed by many insects, and used for placing the eggs; it is a modification of the aculeus.

Ovisac. The external bag in which some of the lower Crustaceans carry their eggs for a time after they are extruded from the ovary.

Ovoviviparous. When eggs are retained until they are hatched.

Ovum, egg, or *germ.* "A highly differentiated portion of the parent organism."

Palæozoic. The primary fossiliferous period in geology; it includes the Laurentian, Cambrian, Silurian, Devonian, Carboniferous, and Permian formations.

Palingenesis. Recapitulative ontogeny.

Pallium. The mantle or fold of integument on each side in the Mollusca.

Palpi. Jointed appendages of the mouth in Arthropoda.

Palpocil. A hair-like process in the tentacles of some Hydroida.

Paluli. Small detached rods round the columella of an Actinozoon.

Pancreas. A conglomerate gland situated behind the stomach and connected with digestion.

Pangenesis. Mr. Darwin's hypothesis that countless "gemmules" are generated in every organ which, having the power of reproduction, are the cause of the appearance in offspring of ancestral characters or of physiological peculiarities.

Panspermism. The hypothesis that all organisms come exclusively from preexisting living germs.

Paraglossæ. Two delicate membranous organs placed behind and above the labium in certain insects.

Parapleuræ. The flanks or sides of the thorax.

Parapodia. The unjointed feet of the Annelida and of the larvæ of many insects.

Parapophyses. The lower articular transverse processes of certain vertebræ.

Paraptera. See *Tegulæ.*

Paratoids. Glandular tubercles, chiefly found above the tympanum in Batrachia.

Parostosis. The fibrous matrix in which integumental bones are developed.

Parthenogenesis. Virginal reproduction. "Asexual reproduction either by fission, gemmation, or the process of internal budding" (*Huxley*).

Patagium. (1) The expanded integument forming the wing of a bat. (2) A vesicular body, one on each side, attached to the prothorax of the Lepidoptera; it is covered with hair.

Paxillæ. Clusters of spines or bristles in the Echinoderms.

Pecten or *Marsupium.* A plicate membrane of the eye, placed in the vitreous humour anterior to the retina.

Pectines. A pair of comb-like organs situated behind the two posterior legs of scorpions.

Pedicellariæ. Small two- or three-pronged pincer-like bodies, found on most Echinoderms.

Pedicelli or "*Ambulacra,*" or "*Ambulacral feet.*" The suckers of Echinoderms.

Pedicle or *Pedicel.* A diminutive peduncle, variously applied.

Pedipalpi. The maxillary palpi of the scorpions, the large prehensile claws.

Peduncle. A foot-stalk, variously applied.

Pelvis. The bony "hip-girdle" supporting the lower extremities.

Pereion. The thoracic segments of the lower Crustaceans.

Pereiopoda. The legs attached to the body (pereion) of the lower Crustaceans.

Periosteum. The fibrous membrane covering a bone.

Periostracum. The membrane covering the shell of the Mollusca.

Perisarc. The chitinous envelope of some Hydrozoa.

Perisome. The calcified integument of an Echinoderm.

Peristome. The part surrounding the mouth of an Actinozoon.

Peritonæum. The serous membrane investing the intestines, and reflected on the walls of the abdomen.

Peritreme. The aperture or mouth of a univalve shell.

Persistent types are those which have not undergone any notable modification in geological time, and even exist at the present day; they belong chiefly to the lower forms of life. There is no "valid reason for the supposition that the earliest forms known in the oldest fossiliferous rocks were the first to make their appearance on our planet" (*Huxley*). Animals with what Mr. Darwin calls "an inflexible organization" may also be referred to this type.

Persona. See *Individual.*

Phalanges. The bones composing the digits.

Pharynx. A membranous sac, in which, *inter alia*, the mouth and the œsophagus open.

Phragmacone. The conical endoskeleton of a Belemnite.

Phylogenesis. The development of the race from the lowest to the highest forms.

Phylogeny. "A branch of biological speculation" which seeks to determine the ancestral history of species.

Phylum. Häckel's name for a subkingdom.

Physoclistous. When the swim-bladder of a fish has no duct.

Pilidium. The helmet-shaped larva of certain worms.

Pineal gland. A small non-glandular body connected to the optic thalami.

Pituitary gland. A two-lobed non-glandular body at the base of the brain.

Placenta. A vascular structure formed by the interlacing of the villi of the chorion and the inner membrane of the uterus in pregnancy.

Placoid. Applied to bony fish-scales, whether in the form of plates, grains, or spines.

Plantigrade. When a Mammal walks on the soles of its hind feet.

Plantula. An accessory joint between the claws of certain insects.

Planula. The locomotive embryo of many Hydroids.

Plasma. The fluid part of the blood in which the corpuscles float. "By the evolution of albuminous granules and oil-particles plasma becomes yolk" (*Owen*).

Plasmodium. A complex network of protoplasm, only observable in the lowest organisms.

Plastide. An independent mass of protoplasm.

Plastron. The ventral plate of Chelonia.

Pleon. The seven abdominal segments or somites of some of the lower Crustaceans.

Pleopoda. The appendages attached to the abdominal segments of some of the lower Crustaceans.

Pleura. The serous membrane investing the lungs and reflected on the walls of the thorax.

Pleurodont. When a tooth is attached to the jaw by one of its fangs with the inside of the socket.

U

Plexus. A network of nerves.

Plica semilunaris. See *Membrana nictitans.*

Plumules. The down-feathers.

Pluteus. The larval form of the Echinoidea.

Pneumatocyst. The air-sac contained in a pneumatophore.

Pneumatophore. The air-sac or float attached to the proximal end of the cœnosarc of the oceanic Hydrozoa.

Podex. The last segment of the abdomen in insects.

Podium. The muscular so-called foot of certain Mollusks.

Polian vesicles. Interradial sacs connected with the ambulacral system of Echinoderms.

Polymely. The monstrosity when one or more of the extremities are doubled, or when there is any supernumerary part.

Polymerism. A multiplicity of parts.

Polypary. The chitinous sheath investing more or less the cœnosarc in the Hydroida.

Polyphyletic. When the descent is from more than one form.

Polypide. The zooid of the Polyzooid colony.

Polypite or *hydranth.* The "nutritive zooid" of the Hydroid colony.

Polyplast. See *Morula.*

Pons Varolii. The commissure of the cerebellum connecting the two lateral lobes.

Primitive groove. A straight shallow depression of the blastoderm, indicating the longitudinal axis of the future embryo.

Procœlous. Said of vertebræ which are concave only in front.

Proglottis. A reproductive segment of a Cestode.

Prognathous. Projecting jaws and teeth; in man a sign of degradation.

Pronotum. The dorsal surface of the prothorax.

Propodite. The sixth joint of the leg of an insect.

Propolis. A substance collected by bees, composed of wax and resin.

Proscolex. The earliest larval stage of a Cestode.

Prosthema. The upright leaf-like process on the nose of certain bats.

Prosthenic. Having the maximum of strength in the fore extremity.

Prostomium. The segment bearing the rudimentary head in the Annelida.

Prothorax. The anterior segment of the thorax in insects.

Protoplasm. "The physical basis of life." An albuminoid substance, homogeneous in the first instance, but capable of assimilation and differentiation. It is a compound of hydrogen, oxygen, nitrogen, and carbon. *Protein* and white-of-egg are very nearly identical with it.

Protopodite. The basal division of the appendages of a segment of a Crustacean.

Proventriculus. The dilated inferior portion of the œsophagus in birds.

Pseudo-hæmal system. In Annelids, canals filled up by a clear red or greenish fluid.

Pseudonavicellæ. Peculiar bodies developed by the Gregarinida.

Pseudonychium. See *Plantula.*

Pseudopodia. Extensions of the sarcode in the Rhizopoda, acting as feet and as organs of prehension.

Pseudova. Eggs of the Aphides, hatched without fecundation.

Pseudovaria. Ovaries which produce gemmules in cases of parthenogenesis.

Pterygoda. See *Tegulæ.*

Pterygoid. The wing-like process on each side of the sphenoid; it is a distinct bone in the lower Mammalia.

Pterylæ. The bands or tracts marking the grouping of feathers.

Pulvillus. A tuft of hairs attached to the underside of the tarsal joints in some insects.

Pupa. That stage in the life of an insect before it assumes the perfect or imago form.

Pygidium. The rump; a general term for the posterior part of the body.

Quadrate bone. A bone which is placed between the upper and lower jaws in reptiles and birds.

Radius. One of the bones of the forearm (*antebrachium*).

Rectrices. The quill-feathers of a bird's tail.

Rediæ. The tadpole-shaped larvæ of the Digeneous worms when they have thrown off their ciliated skin.

Remiges. The quill-feathers of a bird's wing.

Repetition. When there occurs a succession of similar parts in the same animal, as in the centipede.

Retina. The third tunic of the eye, formed by an expansion of the optic nerve.

Retinaculum. . A minute scale or plate which checks the too great protrusion of the sting of certain insects.

Reversion or *Atavism.* When a character reappears in an individual animal which had disappeared for some preceding generations.

Rhachis. The axis of a feather.

Rudimentary organs. Parts which have been gradually atrophied owing to disuse. A rudimentary part is larger comparatively in the embryo than in the adult.

Sacrum. Anchylosed vertebræ, to which, on each side, the pelvic bones are attached.

Sarcoblasts. The "yellow cells" of the Radiolaria.

Sarcode. See *Protoplasm.* The term is most frequently applied to the protoplasm of the Protozoa.

Sarcolemma. "The elastic tunic of the striped muscular fibre."

Scape. (1) The basal joint of the antennæ of insects when unusually developed. (2) The axis or rhachis of a feather.

Schizocœle. The perivisceral cavity that results from the splitting of the mesoblast.

Sclerenchyma. Coral tissue.

Sclerobase. A form of skeleton in the Alcyonaria "formed by the cornification or calcification of the axial connective tissue of the zoanthodeme" (*Huxley*).

Sclerotic. With the cornea the exterior tunic of the eye; in many of the Vertebrata it is ossified.

Scolex. The second larval stage of a Cestode worm when it has encysted itself.

Scrobe. A groove in the rostrum of certain Coleoptera for the reception of the scape.

Scutellum. The posterior part of the mesothorax of insects seen from above. It is variously modified, but in Coleoptera it is triangular.

Scutes. The bony scales of the Crocodilia.

Scyphistoma. A form developed from the hydra-tuba of certain Hydrozoa; it afterwards passes into the strobila stage.

Segmentation or *yolk-division.* One of the changes occurring in the egg after fecundation, by which it becomes divided into cells.

Segments or *somites.* The transverse rings which go to make up the body of an Arthropod or of an Annelid. *Metamere* is used in a more special sense.

Septum. A partition—bony, muscular, membranous.

Sesamoid bones. Small bones developed in the tendons.

Siagonopoda. The two pairs of maxillæ and first pair of maxillipedes in certain Crustacea (*Spence Bate*).

Siphon. The respiratory tube of certain bivalve Mollusca; but it is also used in a more general sense.

Siphunculi. The hollow anal processes of the Aphides.

Somatic cavity. See *Cœloma.*

Somites. See *Segments.*

Species. Individuals having certain characters in common while absent in other individuals.

Spermatoa. "The nucleated cell in which the spermatozoa are developed" (*Owen*).

Spermatozoa. The minute moving flagellate plastides forming the male element. "The nature of the influence of the male element upon the female is wholly unknown" (*Huxley*).

Sphæridia. Minute transparent bodies found in the Echinoidea.

Sphenoid. A bone placed at the base of the skull, articulated as well with the bones of the face.

Spiculum amoris. The love-dart of the snail.

Spiracles. See *Stigmata.*

Sporosac. "A gonophore destitute of obvious umbrella" (*Allman*).

Statoblast. A gemmule enclosed in a peculiar bivalve shell in some of the Polyzoa, agamogenetically developed.

Steatopygons. When there is an unusual development of adipose matter posteriorly.

Stemmata. See *Ocelli.*

Stigma. A small opaque spot on the fore wing of certain insects.

Stigmata. Pores through which air is admitted into the tracheal vessels of insects.

Stipes. A stalk; the basal portion of the maxillæ of insects.

Stolons. In zoology connecting processes of the cœnosarc, &c.

Stomatodendra. The branches of the tree-like mass of the syndendrium of the Rhizostomidæ.

Strobila. A stage in the development of certain Hydrozoa.

Stroma. The bed in which a tissue or organ originates.

Struggle for life. Mr. Darwin uses "this term in a large and metaphorical sense, including dependence of one being on another, and including (which is more important) not only the life of the individual, but success in leaving progeny."

Swim-bladder, air-bladder, or *sound.* An organ filled with air, placed under the spine in fishes. It is the homologue of the lungs.

Symmely. A monstrosity, as when two or more parts are joined together.

Symmorphs. Parts having the same form.

Synangium. The bulbous end of the aortic trunk in Amphibia.

Synapticulæ. "Transverse props" between the septa in a corallite.

Syncytium. The ectoderm of certain Spongia in which the nuclei show no trace of being contained in cells.

Syndendrium. The complex tree-like mass dependent from the umbrella of the Rhizostomidæ.

Synthetic. When a combination of characters occur which normally find their expression in other groups.

Syrinx. The lower larynx, the chief organ of the voice in birds.

Syzygium. An unmovable suture.

Tabulæ. Horizontal plates in the thecæ of certain corallites.

Tarsus. In mammals and some other vertebrates the small bones of the foot; in insects the small consecutive joints ending the legs.

Taxonomy. "The principle of classification."

Tectology. The science of the laws of the grouping of parts which go to make up an individual.

Tectrices or *wing-coverts.* The small feathers on the forearm are the lesser, and those over the quill-feathers are the greater coverts—*tectrices primæ* and *tectrices secundæ* respectively.

Tegmina. The upper wings of the Orthoptera and homopterous Hemiptera; they are uniform in texture, and have no lining membrane beneath.

Tegulæ. Scales, one on each side, attached to the mesothorax of the Lepidoptera.

Teleology. The doctrine of final causes, or, the indication of design and purposiveness.

Telson. The last segment, or the appendage of the last segment, of the abdomen of certain Crustacea.

Tentacula. Prehensile organs of the Cœlenterata, and especially of the Hydroida.

Teratology. The study of abnormal forms or monstrosities.

Terebra. The boring-organ of insects.

Tergal. Relating to the back.

Test. The shell of Mollusca or of Echinoidea; also the tunic of the Ascidians.

Thalerophagous. Feeding on decomposing animal or vegetable matter.

Thaumatogeny. Genesis by miracle.

Thelytocous. When a parthenogenetic female produces only female offspring.

Thermigenous. Generating heat.

Thomia. The sharp edge of a bird's bill.

Thorax. The chest, the upper or anterior part of the trunk.

Tibia. The shin-bone of mammals, in birds the drumstick. In insects it is the fourth joint of the legs.

Tornaria. The larval form of *Balanoglossus.*

Trachea. The wind-pipe. In insects and their allies the *tracheæ* are tubes subservient to respiration.

Trichocysts. See *Nematocysts.*

Trochal disk. The wheel-shaped ciliated organ of a Rotifer.

Trochanter. A process of the upper part of the thigh-bone. In insects it is the second joint of the leg; but it appears to have no separate motion distinct from the femur to which it is attached.

Trochantin. A small movable piece of the exoskeleton of insects placed on the outer part of the coxa.

Trochosphæra. A larval form of Mollusca where the head is girdled with a row of cilia which ultimately becomes the *velum.*

Trophi. The parts of the mouth concerned in feeding.

Trophosome. The "assemblage of zooids" in the Hydrozoa "destined for the nutrition of the colony."

Tympanum. The drum of the ear. In insects certain membranous parts subservient to sound.

Typhlosome. A peculiar fold of the intestine in the Tunicata and the Lamellibranchiata.

Typical. What is the most representative of a group.

Ulna. One of the two bones of the forearm.

Ulotrichous. Having woolly hair; the hair is also flattened, and its section oval.

Umbo. The boss or beak near the hinge of a bivalve shell.

Umbrella. "The gelatinous bell of a medusiform planoblast" (*Allman*). "A swimming-bell with the velum" (*id.*). Without a velum (*Huxley*).

Unguligrade. Walking on the hoofs.

Uropoda. The three pairs of expanded hind legs in certain Crustacea.

Uropoietic system. The organs for the secretion of water and urea ; in the invertebrates they are supposed to be represented by the water-vascular system.

Urosthenic. When the maximum of strength is in the tail.

Urostyle. A prolongation of the spinal cord in certain fishes and amphibians.

Uterus. The womb.

Vacuoles. Certain cavities in the Rhizopoda having a contractile or rhythmical movement.

Variability. All individuals vary, but accidental variability is due to "indeterminate antecedents," and only exists to a "very small extent," and only in plastic forms. Some species have a "singularly inflexible organization" (*Darwin*).

Variety. Any departure from the parental type.

Veliger. An advanced larval form of Mollusca when the velum is fully developed.

Velum. (1) In a medusiform gonophore, "the membranous perforated diaphragm which stretches transversely across the codonostome" (*Allman*). (2) A ciliary expansion of the integument attached to the head in the larval Mollusca.

Vena cava. The great vein that returns the blood to the heart.

Ventricle. A cavity; one or two in the heart, two in the brain.

Vertebræ. The bones forming the spinal column.

Vestibule. A cavity within the ear.

Vexillum. The web, composed of barbs, with its scape or rhachis, of a feather.

Vibracula. A peculiar cup-shaped appendage in the Polyzoa, to which is attached a movable seta.

Vibrissæ. Stiff hairs ; a general term.

Viscera. Internal organs, especially those within the thorax and abdomen.

Vitellus. The yolk of an egg.

Vomer. A bone forming part of the septum of the nares, and connected more or less with the palatine bones.

Wallace's line. A line which is assumed to divide the "Malayan" from the "Austro-Malayan" regions. It passes between Bali and Lombok in the south, through the Macassar Straits, dividing Borneo from Celebes, and to the north-east between Mindanao and Gilolo.

Yellow cells or *sarcoblasts.* Peculiar nucleated structures in the Radiolaria, containing yellow protoplasm (possibly parasites).

Zoanthodeme. The body formed by the coherence of many zooids of a single polyp of an Actinozoon.

Zoarium. See *Cœnœcium.*

Zoëa. An intermediate larval stage in the higher Crustaceans.

Zonites, Somites, or *Metameres.* See *Segments.*

Zoœctum. A cell in which a polypide of the Polyzoa is lodged.

Zooid. "A term applied to the individuals of compound organisms, as the polyps of a polypidom" (*Huxley*). "The detached portions of an individual in discontinuous development" (*Greene*). "The more or less independent products of non-sexual reproduction" (*Allman*). *Zooid* is a name given also to the central basis of a blood-corpuscle.

Zoology. The science of animals.

Zoomorphism. The series of changes in the life of an animal.

Zoon. Mr. Herbert Spencer's name for "the whole product of a fertilized germ."

Zootheme. The compound animal mass produced by budding.

Zygapophyses. Certain processes of the vertebræ.

Zygosis. See *Conjugation.*

INDEX

Paradoxurus

Y

THE END.

www.ingramcontent.com/pod-product-compliance
Lightning Source LLC
Chambersburg PA
CBHW021458210326
41599CB00012B/1051